PREDISPOSED

LIBERALS, CONSERVATIVES, AND
THE BIOLOGY OF POLITICAL DIFFERENCES

BY
JOHN R. HIBBING
KEVIN B. SMITH
JOHN R. ALFORD

Routledge
Taylor & Francis Group

NEW YORK AND LONDON

First published 2014
by Routledge
711 Third Avenue, New York, NY 10017

Simultaneously published in the UK
by Routledge
2 Park Square, Milton Park, Abingdon
Oxon OX14 4RN

Routledge is an imprint of the Taylor & Francis Group, an informa business

Library of Congress Cataloging in Publication Data

Hibbing, John R.
Predisposed : liberals, conservatives, and the biology of political differences / by John R. Hibbing, Kevin B. Smith, John R. Alford.
    pages cm
  1. Political culture.   2. Political sociology.   3. Political participation.   4. Liberalism.   5. Conservatism.
I. Smith, Kevin B., 1963–   II. Alford, John R.   III. Title.
    JA75.7.H53 2013
    320.50973—dc23
    2013013341

ISBN: 978-0-415-53587-8 (hbk)
ISBN: 978-0-203-11213-7 (ebk)

Typeset in Minion
by Apex CoVantage, LLC

Printed and bound in the United States of America by Sheridan Books, Inc. (a Sheridan Group Company).

To Anne, Kelly, and Mendy

CONTENTS

ACKNOWLEDGMENTS

We wrote a popular rather than an academic book about political predispositions for a reason: We think it is important for a wide range of people to understand why not everyone sees the world the same way they do. Recent scholarly research on biology's connection to political temperament might increase political understanding by making it easier to deal with political differences and conflict, but these studies are not readily accessible to a general audience. Reports published in professional academic journals often include eye-glazing technicalities and, when the results of this research are picked up by the popular press, much is lost in translation. We appreciate the media making research accessible to a broad audience, but the often-brief summaries tend to leave a false impression of the central findings. Our aim is to provide a book-length treatment that summarizes the recent research in a detailed but readable fashion. Our message is not one that everyone will like, but please don't kill the messenger. We did not construct the complex, fascinating, flawed, and infuriating creatures called humans; we just study them. By understanding the human condition, warts and all, we think it might be possible to build and maintain political systems that work better than they currently do. We know for sure that pretending that humans are something they are not will only lead to frustration and further polarization of the political arena.

We are eager to acknowledge the valuable assistance we received in writing this book and in conducting the research that made it possible. First, primary financial support was provided by the National Science Foundation in the form of grant BCS-0826828. Second, our students throughout the years have shaped our thinking and improved our research dramatically. We were fortunate that during the drafting of this book we had an unusually talented group of graduate

students, including Tim Collins, Kristen Deppe, Balazs Feher, Amanda Frei-
sen, Karl Giuseffi, Frank Gonzalez, Mike Gruszczynski, Carly Jacobs, Jayme
Neiman, John Peterson, and Ben Seiffert. Jayme comes in for special thanks.
She was absolutely indispensable in many roles: reading the entire manuscript,
offering insightful suggestions, designing figures, and tracking down fugitive
bibliographic and other information. Colleagues and administrators at our
respective universities, and in particular in the Department of Political Science
at the University of Nebraska–Lincoln, have been wonderfully supportive. Mat-
thew Hibbing and Jessica Mohatt provided crucial feedback on earlier drafts
and saved us from many errors. We thank our colleagues around the world, in
a variety of disciplines, who are also engaged in investigations of the sources of
political attitudes, sometimes in collaboration with us and sometimes independently. We enjoy being their fellow travelers. We thank our editor at Routledge,
Michael Kerns, for his support, encouragement, and guidance on this project,
as well as his excellent taste in wine. And we thank our agent, Judy Heiblum,
for being such an unflagging advocate for this project despite the fact that it
happened when she was busy moving to a different continent and having a
baby. Mostly, we thank our spouses, a couple of whom (Anne Nielsen Hibbing
and Kelly Smith) were dragooned into reading the entire book and providing
critiques, support, and suggestions.

BIOGRAPHICAL STATEMENT

John R. Hibbing is the Foundation Regents University Professor of Political Science at the University of Nebraska–Lincoln; Kevin B. Smith is Professor of Political Science at the University of Nebraska–Lincoln; and John R. Alford is Associate Professor of Political Science at Rice University. Together, for the past decade, they have been investigating the biological and deep psychological bases of political orientations. Their research has appeared in leading academic journals, including *Science, Behavioral and Brain Sciences,* and the *American Political Science Review,* and has been featured in hundreds of stories and segments in popular media outlets.

CHAPTER 1

Living with the Enemy

Democrats: Sweaty, disorderly, offhand, imaginative, tolerant, skillful at give-and-take.
Republicans: Respectable, sober, purposeful, self-righteous, cut-and-dried, boring.

Clinton Rossiter, *Parties and Politics in America*

Politics is a blood sport where fights among spectators can be just as ferocious as the blows traded by combatants. Political exchange tends toward the emotional and primal rather than the reasoned and analytical, which is why it must have seemed like a good idea to ABC News in 1968 to televise a series of debates between William F. Buckley, Jr., and Gore Vidal. Both were ideologues—Buckley for the right, Vidal for the left—but ideologues in an educated, patrician, and articulate men-of-letters sort of way. Perhaps they could demonstrate to a mass audience that it was possible for debates between political opponents to employ words that were honest, intellectual, and constructive rather than pejorative, dismissive, and rancorous.

That sort of example was desperately needed in the United States in 1968, a time when people who disagreed with the political ideas of other people had picked up an alarming habit of shooting them or beating them senseless. Robert Kennedy and Martin Luther King, Jr. were assassinated; race riots raged in dozens of cities; and during the Democratic National Convention, anti-Vietnam War protestors fought the Chicago police for control of the city's streets in an epic eight-day running battle. Buckley and Vidal, then, must have seemed like just the ticket. They were smart and hyper-articulate, and their plummy, East Coast establishment tones made them seem so, well, civilized. Perhaps they could demonstrate a more mature way to deal with political differences. Or not.

In their most famous exchange, on August 27, 1968, Buckley asserted that Vidal was unqualified to say anything at all about politics, calling him "nothing more than a literary producer of perverted Hollywood-minded prose." Vidal retorted that Buckley "was always to the right, and always in the wrong," and accused him of imposing his "rather bloodthirsty neuroses on a political campaign."

After that the gloves came off.

"Shut up a minute," said Vidal. Buckley did not shut up. Vidal called him a "proto- or crypto-Nazi." Buckley was not happy with that. "Now listen you queer," he said. "Stop calling me a crypto-Nazi or I'll sock you in the goddam face."[1] Buckley went home in a huff and sued Vidal for libel. Vidal went home in a huff and, perhaps miffed that he didn't think of it first, counter-sued Buckley for libel.

So much for a civilized exchange of views.

At this point we could cluck our tongues and make highbrow academic noises about the degeneration of political exchange. We could point back to the early days of the American experiment and hold up the dignified Founders as better examples of civil and edifying political debate. We won't, though, because they, too, stuck in the shiv when it suited them. Like Buckley and Vidal, Alexander Hamilton and John Adams could be insufferable know-it-alls, intolerant of viewpoints other than their own. President Adams signed into law the Alien and Sedition Acts, making it a crime to say nasty things about the government—a good deal if you are head of that government—and Hamilton engaged in a personal feud with Vice President Aaron Burr so vitriolic it ended with Burr putting a musket ball through him. As an example of politics putting people on the boil, it is difficult to top the sitting vice president of the United States shooting and killing one of the prime movers and signatories of the Constitution. Other Founders weren't much better. Thomas Jefferson and James Madison, held up in the United States as semi-divine political angels descended from Mount Independence, chucked mud with the best of them. Jefferson, for example, slandered his opponents on the sly. He bankrolled

James Callender, a professional "scandal monger," to attack Adams. Callender obliged, describing Adams in scorching prose as "a repulsive pedant, a gross hypocrite and unprincipled oppressor."[2] Callender was tossed in jail for violating the Alien and Sedition Act (score one for Adams) and Jefferson got a nasty bit of blowback—he fell out with his journalistic attack dog, who promptly turned to writing titillating tales about Jefferson's affairs with an attractive slave named Sally Hemmings. Jefferson indignantly and, if you believe DNA testing on Hemmings' descendants, wholly misleadingly said he did not have sexual relations with that woman.

Don't be smug if you are not from the United States; we're willing to bet your political icons are not much different from the feet-of-clay rhetorical flame-throwers blistering each other under the Stars and Stripes. Show us a paragon of politics from any time and place and chances are we won't have to scratch the surface too hard before finding something like the Buckley-Vidal kerfuffle, in other words someone saying the other guy's political views are so wrongheaded they merit a fast-moving fist to the schnoz.

People take politics seriously. They love validation of their own opinions and vilification of their opponents' opinions. This is why conservatives make Ann Coulter, Michelle Malkin, and Mark Levin best-selling authors, Rush Limbaugh a wealthy talk-radio titan, and Fox News the most watched outlet on cable television. These sources can be counted on to tell their audiences that conservatives are noble defenders of the good and the just while liberals are stubbornly mugger-headed and oppositional. Driven by a desire to receive precisely the opposite message, liberals flock to the books of Al Franken, Michael Moore, and Molly Ivins, and the satire of television comedy like *The Daily Show with Jon Stewart*. Stephen Colbert of *The Colbert Report* has created a massively successful career around the persona of a shallow, jingoistic, uniformed conservative buffoon. Diatribes against liberals or conservatives enjoy a guaranteed audience of partisans all subscribing to the maxim, "why be informed when you can be affirmed."[3]

If we were of an avaricious bent, we could write another broadside against stupid, inbred, uninformed, malodorous, bloodsucking conservatives. If we really wanted the big bucks, we could pen a blistering condemnation of duplicitous, malevolent, degenerate, cretinous liberals. Such works sell very well among certain demographics and, having read a fair sampling of what's on offer, we see little evidence that it takes much effort or talent to get on a good rant. Authors of these popular political screeds rarely seem to invoke—let alone conduct—systematic research. Ginning up a truckload of demeaning adjectives to unload on one group or another? Sounds like it might be fun as well as profitable. Unfortunately, we are academics, so neither profit nor fun is what interests us most.

Besides that, the world does not need another book assuring readers that their political views are laudably correct while those of their political opponents are pathetically, dangerously, and rashly incorrect. Such books only pander to the worst instincts of those who care deeply about politics, encouraging extremity and discouraging dialogue. Ad hominem attacks on the political "other side" may be comfortingly confirmatory to readers and financially fulfilling to authors, but they are shallow, derivative, and polarizing.

In this book we aim to explain why people experience and interpret the political world so very differently. We want to provide liberals with a better appreciation for the conservative mindset; conservatives with a better appreciation for the liberal mindset; and moderates with a better appreciation for why those closer to the extremes make such a big fuss. We make no pretense that conservatives and liberals can be led to agree on everything, or even anything. Getting the Buckleys and the Vidals of the world to hold hands and sing "Kumbaya" around a campfire is just not going to happen. Pretending that some middle-ground nirvana can be reached if only we listen to the other side is counterproductive and a source of endless frustration. We are after smaller but important and much more realistic game. We want liberals and conservatives to understand why they are different from each other and why those differences frequently seem so unbridgeable.

We recognize what we are up against. Liberals and conservatives are rarely in the mood to understand the other side. This resistance to accepting the other side is something we have encountered in our own professional lives. A few years ago, we published a study showing that liberals and conservatives experience the world differently and suggested that it might be unproductive and slightly inaccurate to view either side as irredeemably malevolent—or unremittingly beneficent. Media coverage of this study led to us to receive numerous emails. Some of these were decidedly caustic, but the most memorable was more plaintive than judgmental. Its key line was "don't do this to me: I NEED to hate conservatives." Clearly, for some it is deeply rewarding to denounce political adversaries, preferably at high volume.

Facing Your Monsters

"If you're not outraged, you're not paying attention," goes the old saying. We disagree. Outrage does not solve challenging issues of governance and it is possible for people to pay close attention to politics without losing emotional control. A more productive, if less viscerally satisfying, response to political difference is to try to understand the source of the views of those who disagree with us so fundamentally. Doing so does not mean your resolve is weakening or that your fellow travelers should begin to worry about you; making an honest effort to understand the other side is not selling out.

In urging each side in the political debate to work harder at accepting the other side, we are not implying that the two poles of the ideological spectrum are mirror images of each other and equally culpable on all matters. The media often engages in "false equivalency," leaving the impression that if a problem exists, both sides must have contributed equally to its creation. For example, if one side of the political debate is not compromising then the other side must not be compromising either. This is not our position at all. Our pitch is that liberals and conservatives and everyone in between have different orientations to information search and problem solving and therefore contribute to political difficulties and solutions in very different ways. Indeed, one manifestation of this is that the two ideological poles have quite different attitudes toward compromise.

To illustrate the value of entering the mindset of the other side, consider the following. One of our children was given to horrible nightmares. He would cry and shout as monsters circled in his sleep. Words from the awake world ("there is no monster under your bed") could not disabuse him of the fears that were so real to him. Weeks into this unpleasant pattern, due more to desperation than inspiration, his parents' strategy changed. Instead of telling him how silly and outrageous he was being, they entered his dream world. "Yes, there is a monster. Oh, he's an ugly one—mean, too—and he's coming this way. But wait, he just spotted some monster friends of his and he's moving off in another direction— way off." By imagining the world he was in and by letting him know that others understood the nature of that world, it became possible to work through the attending issues. Blissful sleep—for parents and child alike—soon descended where monsters had lurked only moments before.

Dismissing the nightmare world of political adversaries is a wholly ineffective approach to solving political problems. What is lost by making a real effort to enter their world, not with the intention of joining them but to understand the reasons they have come to such different political conclusions? You are free to believe that the world of your political adversaries is as detached from reality as a scared little boy's nightmare world—but realize it is as real to them as the monsters were to him. Also realize that to your political adversaries, your world is as detached from reality as a child's green, scaly monster. Maybe if we understand their world we can figure out how to live with people who annoyingly, irritatingly, and persistently come to political viewpoints so very different from our own. Puzzlement is better than hate.

In this book we make the case that political variations are part of an incredible range of differences in the way people respond to the world. Just to give you a brief teaser, it turns out that liberals and conservatives have different tastes not just in politics, but in art, humor, food, life accoutrements, and leisure pursuits; they differ in how they collect information, how they think, and how they view other people and events; they have different neural architecture and display distinct brain waves in certain circumstances; they have different personalities and psychological tendencies; they differ in what their autonomic nervous systems are attuned to; they are aroused by and pay attention to different stimuli; and they might even be different genetically. At least at the far ends of the ideological spectrum, liberals and conservatives are emotionally, preferentially, psychologically, and biologically distinct. This account is not just based on casual observation or armchair analysis. Science—both social and biological—is our co-pilot.

Liberals and conservatives often are reluctant to accept that their differences are rooted in psychology, let alone biology. Their own political beliefs seem so sensible, rational, and correct that they have difficulty believing that other people, if given full information and protected from nefarious and artificial influences, would arrive at different beliefs. Liberals are convinced the existence of conservatives can be written off to Karl Rove's treachery, the Koch brothers' fortune, the bromides of Fox News, and a puzzling proclivity to think simplistically. Conservatives are equally convinced the existence of liberals is attributable to the "lamestream" media, indoctrination by socialist university professors, the sway of Hollywood, and a maddening tendency to disengage from the real world. Yet political differences are grounded not in a duplicitous conspiracy or an irrational disregard of logic and truth but rather in variations in our core beings. Conservatives are not duped liberals and liberals are not lazily uninformed conservatives.

You would not come to this conclusion by looking at much of today's popular political commentary. Egged on by ideologically biased authors and personalities, efforts to understand political opponents often go no further than the assertion that they are ignorant, obdurate, and uninformed—those on the right are "big fat idiots" and those on the left are "pinheads."[4] Accepting that political differences are due not merely to incorrect information, elite machinations, or an unwillingness to think through situations is an important step toward living more comfortably. A better understanding of the biological and psychological realities of our political opponents makes it possible to recognize that their policy preferences, however misguided to our eyes, are sensible given their different realities. Getting to that point is crucial. As journalist Robert Haston put it, "[W]e can accept and understand the red or blue tribal instincts that drive the other half, or we can continue our retreat into ever more blind and vicious combat."[5]

Nobody's Perfect, but We're Working on It

We are often asked why we research the deeper bases of political differences and invariably our questioners assume that the real goal must be to paint one political group or another in an unfavorable light. We must be a bunch of academic lefties trying to stick it to the right. Or maybe we are traitors to the cause and are out to disparage the left. The notion that social scientists might be studying the nature of the human condition without promoting an alternative agenda is rarely accepted, particularly when the topic is politics.

The central message of this book is that lurking predispositions are widespread, so we would be the last people to claim social scientists or anyone else can be 100 percent objective and value free. If you think you are not biased, you are fooling yourself. You get an exception only if you have pointy ears, green blood, and a commission from Star Fleet. While we are just as biased as every-one else, the great thing about the scientific process is that the biases of a single research team eventually get squeezed out. In our bailiwick, data and evidence ultimately rule, or at a minimum have more influence than hidden political agendas. Our world is the world of empirical social science, a pretty ruthlessly Darwinian piece of real estate. It revolves around an ongoing scientific process that affords skeptics the chance to participate fully. Different researchers weigh in, replication will occur (or not), and eventually the truth will emerge—not a definitive or ultimate truth but the best current shot at the (relatively) unbiased truth. You should be on guard for suspect methods and biased inference but you should not be paralyzed by suspicion. You should be skeptical of the results of a single study, including anything we have published. Yet if numerous studies conducted by numerous labs with alternative techniques and in diverse settings begin to point in the same direction, you should accept that the burden of proof shifts to those who deny that liberals and conservatives have deep differences.

Unfortunately, when it comes to politics, the distinction between system-atic, validated description and howling ridicule is all too often ignored, the upshot being that any research showing psychological or biological differences between liberals and conservatives is reflexively treated by one side or the other, and often both, as biased. To take one example, evidence that conservatives are more conscientious while liberals are more tolerant of uncertainty is thought to be less an effort to understand political temperament than an attempt to belittle liberals/conservatives (take your pick) while hiding behind the veneer of science. When it comes to ideology, difference equals value judgment in the minds of many, when in reality it is possible to be different without being better or worse.

That said, we freely recognize that suspicions of political judgments hiding within social science research (or even in the headlines) are not without foun-dation. One piece of early research on the deeper bases of political attitudes concludes that conservatism is characteristic of "social isolates, of people who think poorly of themselves, who suffer personal disgruntlement and frustra-tion, who are submissive, timid and wanting in confidence, who lack a clear sense of direction and purpose, who are uncertain about their values, and who are generally bewildered by the alarming task of having to thread their way through a society which seems to them too complex to fathom."[6] More recent research describes conservatives as "easily victimized, easily offended, indecisive, fearful, rigid, inhibited, relatively over-controlled, and vulnerable."[7] It is a wonder conservatives can get themselves out of bed in the morning.

Were these conclusions unduly biased? We can say that the two studies cited above were quickly and robustly challenged. Like others who deconstructed the empirical and conceptual case behind those statements, we are skeptical of some of the supporting evidence. That is what the scientific process does. Identifying problems makes it possible to correct them subsequently. The account we present in this book is based not on a single study but on a massive collection of studies conducted by many scholars in many countries. This does not mean the final truth has been discovered; it means that the weight of evidence permits confidence in the claim that liberals and conservatives really are fundamentally different.

We are not trying to demonstrate that conservatives are crypto-Nazis or that liberals are naïfs who need a good sock in the mouth to jar them into recognizing reality. We just want to know why people are so different politically. So, if only for the time it takes to read this book, we ask readers to suspend any instincts to dismiss as crassly biased any research that does not conclude that their political foes are evil incarnate. We will note some imperfections of those foes along the way, but keep in mind that nobody is perfect and the imperfections of liberals are very different from the imperfections of conservatives. The task we set ourselves here is not to tally the imperfections of each ideological group in order to declare one group the winner. We just want to know why the groups exist in the first place.

Whether the topic is climate change, evolution, genetically modified foods, or the biological basis of political beliefs, people are quick these days to apply the label of junk science to research on controversial matters. The implication is that some research is driven by special interests and hidden agendas to such an extent that it cannot be considered real science or, more likely, that some topics are simply not suitable for science. Replication should take care of the hidden agenda issue and as far as some topics not being amenable to the scientific process, consider this. Researchers recently presented one group of people with scientific evidence that confirmed their prior beliefs while a second group received the same evidence but it disconfirmed their prior beliefs. Compared to those receiving belief-confirming evidence, those receiving the belief-disconfirming (but very same) scientific evidence were much more likely to conclude that the topic could not be studied scientifically.[8] In other words, the charge of junk science appears to be nothing more than a lazy way of saying "I don't like the findings."

What about Me? I'm a Libercontrarian

What about those who do not feel comfortable being pigeonholed as liberal or conservative? What about all those folks who live in countries where those two words do not hold much meaning, even when translated? What about all the people who could not care less about politics? A common mistake in addressing differences in political orientations is to leave the impression that they begin and end with the distinction between liberals and conservatives or between those on the political left and those on the right (we use these pairs of terms inter-changeably). These labels are short, convenient, and convey an intuitive notion of political differences. We will use them for exactly that reason throughout this book. Still, it is quite true that they fail to capture the political views of a goodly percentage of people. So before going any further we wish to make it clear that, even though we often use phrases such as liberal and conservative or left and right for shorthand, this book is about political differences generally and not merely differences between two discrete collections of ideological beliefs.

The differences we are talking about reflect variation across a continuum and perhaps many continuums, not traits that lump everyone into two camps. When we say left/right or liberal/conservative, what we have in mind is more a yardstick than two measuring cups. Some scholars think ideology is such a complicated and nuanced critter that it demands more than one type of measurement—sort of like body mass is measured by height and weight, ideology should be measured by, say, views on economic policies and views on social policies. Even so, the unidimensional concept of making sense of political differences captures a very long tradition of describing political differences.

Plato (liberal) and Protagonus (conservative) are sometimes viewed as the progenitors of these political types, though undoubtedly prehistory is chock full of earlier illustrations; Catherine the Great's Russia, for example, was fraught with conflicts over abolishing serfdom, the role of religious freedom, efforts to rein in the nobles, and appropriate attitudes toward the new ideas of the Enlightenment. Nineteenth-century philosopher John Stuart Mill called it "commonplace" to have "a party of order or stability and a party of progress or reform."[9] Ralph Waldo Emerson noted that "the two parties which divide the state, the party of conservatism and that of innovation, are very old, and have disputed the possession of the world ever since it was made." Emerson called this division "primal" and argued that "such an irreconcilable antagonism, of course, must have a correspondent depth of seat in the human condition."[10] That pretty much sums up what we are interested in doing—looking into the depth of the human condition to figure out the irreconcilable antagonism between political beliefs. Capturing that irreconcilable antagonism by distinguishing between competing sets of preferences labeled conservative/right or liberal/left does not do justice to the full range of political preferences people hold, but this distinction has proven a robust way of categorizing the political divisions present in virtually all politically free countries.[11] If it was good enough for Mill and Emerson, it is good enough for us; we'll explain exactly why in the next chapter.

Using these labels, though, could create confusion, and we want to head that off if we can. For example, in some countries a "liberal" refers to a libertarian, a set of political beliefs generally associated with the conservative end of the political spectrum. As a result, in Australia political conservatives belong to the Liberal Party, which may seem a contradiction. In America the best-known libertarians (e.g., Ron Paul) are found in the Republican Party, which champions the distinctly nonlibertarian policies of government involvement in people's social lives (for example, preventing women from having abortions and gay people from marrying the spouse of their choice) and massive levels of spending on defense. Instead of using phrases such as liberal and conservative we could more accurately capture political differences by talking about individual preferences that reflect, for example, a desire for security/protection, a desire for predictability/order/certainty, a desire for equality, or a desire for novel structures and events. This approach would provide some useful flexibility and clarify some of the translation issues that can come with left/right or conservative/liberal labels; unfortunately, it also is incredibly cumbersome. So for ease of communication we will stick with "liberal" and "conservative." We want to make clear, though, that the deeper forces we explore do not demand that there be just two categories of political person. If looked at carefully enough, pretty much everyone's politics are as unique as their physiologies and cognitive tendencies.[12]

Indeed, this potential to account for numerous, diverse political orientations means our story is not just about two political camps in the United States. Our claims apply to other countries and other times. If Mill and Emerson are as correct as we believe them to be, a broad left-right dimension anchors politics universally, even if unique issues and varying collections of positions provide plenty of variation. In sum, our results and interpretations apply to those in the United States who are not comfortable categorizing themselves as either liberals or conservatives and also to those living in countries in which the liberal-conservative distinction is not used. People who are moderates (and there are a good many moderates), libertarians, or Social Democrats are likely to have their own politically relevant predispositions.

Fat Men Can't Jump: Thinking Probabilistically

Accounting for variation in political attitudes is not easy, so if you want to keep that prefrontal cortex in neutral you're in the wrong book; bail now and pick up a copy of Kenny Conservative's *Liberals Blow* or Linda Liberal's *Conservatives Suck* a few shelves down. The social world is messy and full of idiosyncrasies, and making sense of what explains variation in political outlook is going to be a hunt for hints in the gray, not the black and white. Social scientists have no equivalents to the law of gravity or $E = MC^2$. We spend most of our time hunting for modest patterns buried amid remarkable complexity. That is the world we are inviting you into.

Many are skeptical of this world, and not without reason. Whenever a study claims to find something that systematically varies with political orientations, lots of people start thinking of exceptions. For example, at least in the United States, more education is generally associated with more liberal-leaning political preferences. Yet it is easy to cite examples of highly educated conservatives (William Buckley was a Yale alum and conservative columnist George Will has degrees from Oxford and Princeton). Higher levels of religiosity are generally associated with being conservative. Yet there are plenty of pious liberals wandering about (Reinhold Niebuhr—one of the best known twentieth-century theologians—was a committed Christian and also an influential left-winger).

These contrary cases, though, should be kept in perspective. The occasional exception does not negate a pattern. If it is cold today, that does not mean the global climate is cooling. Knowing a lifelong smoker who still runs marathons does not alter the fact that smoking is a serious health risk. Thinking probabilistically rather than deterministically is absolutely key to understanding the message of this book.

All the relationships we describe are only tendencies, not hard and fast rules. Predispositions are not destiny, but defaults—defaults that can be and frequently are overridden. There's a reason the title of this book is *Predisposed* and not *Fated*. But the fact that there is any predisposition at all is important as it tilts subsequent attitudes and behavior in one direction or the other. A person with a particular set of physiological and cognitive traits will not automatically be a liberal or a conservative, but is more likely to be one or the other.

With regard to our approach in this book, we'd like to put our cards on the table. We have a pair of nines. A reasonable hand for five-card stud but not a sure-fire winner. We may not be doing ourselves any favors by confessing that we cannot claim to have discovered the definitive basis of political differences. Nobody likes caveats hanging from their bumper sticker certainties. But we think that much of the skepticism surrounding this line of research stems from people perceiving that the results and claims are stronger than they are. Critics of research on political predispositions are eager to create a straw man by arguing that proponents are making powerful assertions such as "people's political orientations are hardwired from birth" even though those doing the research recoil from those sorts of deterministic pronouncements.

So, as you ponder the message of this book, we ask that you banish "determine" from your vocabulary and replace it with words such as "shape," "influence," "mold," and "incline," and that you be ready for violations of any rule proffered. This is important because, particularly when biological variables are involved, some people tend to think one exception to the claimed pattern negates the entire enterprise. This simply is not the case; biology (and certainly psychology) largely works probabilistically rather than deterministically so exceptions are always to be expected. Eating lots of junk food, for example, increases your chances of suffering from a whole range of health problems. It does not guarantee those problems will actually appear—some candy-snacking fast food devotees stay in good health, the lucky so-and-sos—but it does make it a lot more likely.

To get accustomed to thinking probabilistically, we need a good, simple example. Consider the relationship between a personality trait—for example, conscientiousness—and ideology. Higher levels of conscientiousness correlate with being more conservative, a relationship replicated in a number of independent studies. Fair enough, but what exactly does being "correlated" mean, and how can we vest any confidence that this relationship is real? To begin, we need reliable measures of both conservatism and conscientiousness. Though we can observe indicators of conservatism (say, who you vote for) or conscientiousness (say, whether or not you jaywalk), these are mostly psychological concepts. How conservative or conscientious we are is something that exists mostly in our heads, making measurement challenging. We lack a skull-penetrating measurement machine that tells us how long your conservatism is in milli-cons or what your conscientiousness weighs in consc-o-grams. Social scientists overcome this problem mostly by asking people how conservative and how conscientious they are.[13]

Luckily for present purposes, one of our data sets—taken from a sample of about 340 American adults in the summer of 2010—has a measure of conservatism as well as a measure of personality traits, including a measure of conscientiousness. If we divide our sample into conservatives and liberals and those who are and are not conscientious, we get the distribution displayed in the table in the top panel of Figure 1.1. This shows that 52 percent of conservatives are conscientious compared to 40 percent of liberals. If we reach into this distribution and randomly collar one of our conservative research subjects, our best bet is that he or she will be conscientious. Randomly selecting a liberal, on the other hand, would yield someone conscientious only an estimated 40 percent of the time. There is no certainty to this outcome—only a set of odds that make the conservative research subject marginally more likely to be conscientious and the liberal research subject marginally less likely to be conscientious. In casino terms, a conscientious conservative is the safe bet—and while it will not always pay off, over the long run it will. This general description applies to most all relationships in the social and biological sciences. Certainty is rarely apparent; get used to exceptions.

Though getting across the basic notion of probabilistic relationships, frequency comparisons are pretty crude and uninformative. For one thing, we just completely ignored the main point of the previous section; people's

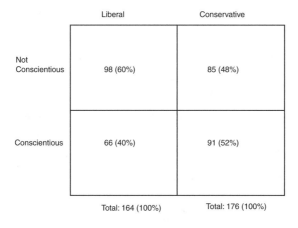

	Liberal	Conservative
Not Conscientious	98 (60%)	85 (48%)
Conscientious	66 (40%)	91 (52%)
	Total: 164 (100%)	Total: 176 (100%)

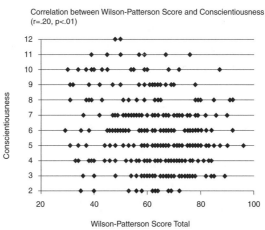

Correlation between Wilson-Patterson Score and Conscientiousness (r=.20, p<.01)

Figure 1.1   Two Ways of Documenting a Relationship

political beliefs do not fit neatly into two distinct boxes, but range across an infinite set of possibilities between the right and left. It is the same deal with conscientiousness—differences with regard to this trait are generally of degree rather than kind. A more accurate approach to assessing these sorts of relationships is through the statistical concept of correlation, which makes it possible to look at measures that have many increments, not just two.

Take a look at the graph in the second panel of Figure 1.1, the one that looks like somebody let fly with both barrels of a shotgun into a barn door. This contains the same information as the table in the top panel. The difference is that the picture—known as a scatterplot—includes all the variation in our measures rather than divvying it up into four liberal/conservative and conscientious/not conscientious boxes. Our measure of conservatism was not a simple are-you-or-aren't-you question. We asked people their opinions—whether they strongly agreed, agreed, disagreed, or strongly disagreed—with 20 issue positions on everything from defense spending to gay marriage.

We converted conservative positions (e.g., strongly disagreeing with gay marriage, strongly agreeing with school prayer) into higher numbers then added together all the scores. This gives us a potential range of 20 (very liberal on all issues) to 100 (very conservative on all issues) with a full range of intermediate positions in between. The liberal/conservative items constitute what is known as a Wilson-Patterson index of conservatism, a set of questions that captures left-right political differences on a wide range of issues, making it possible to measure political orientations as a range rather than just as a category. For conscientiousness we used two questions from a standard "Big Five" test of personality traits. These items asked people the extent to which they saw themselves as "disorderly and careless" and how accurately they felt the statement "I can't relax until I have everything done that I planned to do that day" described them. Responses to the first question ranged from strongly agree (1) to strongly disagree (7); responses to the second ranged from very accurate (1) to very inaccurate (5). Adding these together gives an index with a theoretical range of 2 to 12, with those scoring higher being presumably more orderly, careful, and committed to finishing their to do list every day. In other words, people who are more conscientious.

Great. We computed two measures, plotted them on a standard X-Y axis graph and ended up with something that looks like an aerial shot of Trafalgar Square after a particularly nasty outbreak of pigeon diarrhea. What good does that do us? The answer might be made a little clearer by looking at the scatterplots in Figure 1.2. These plot body weight first with high jump performance (top panel), then shot put performance (middle panel), and then number of nephews and nieces (bottom panel), for 20 hypothetical male athletes. The top two panels show clear relationships; the bottom panel doesn't show any relationship at all. That is about as far as "eyeballing" the data can get you. What about statistics?

All of these visual relationships can be assessed more systematically by something called a coefficient of correlation (often referred to as an *r*), a computation

Correlation between body weight and high jump (r=-.827, p<.01)

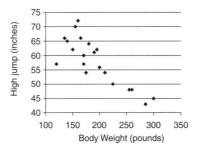

Correlation between body weight and shot-put (r=-.737, p<.01)

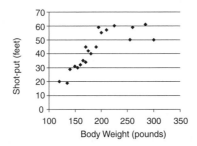

Correlation between body weight and number of nieces and nephews (r=-.019, p>.05)

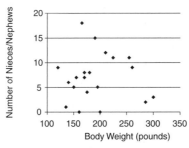

Figure 1.2   Illustration of Negative, Positive, and No Relationship

that spits out a number between −1 and +1 summarizing the extent to which two variables are related to each other. A negative relationship indicates that as one variable goes up, the other goes down. The correlation for the two variables in the top panel is −0.83—in other words, the more you weigh the lower your ability to launch yourself any considerable distance from the ground. The correlation for the two variables in the middle panel is 0.74—in other words, bigger guys can chuck cannonballs farther. These are really high correlations and they also make sense. It is difficult to imagine sumo wrestlers Fosbury Flopping over six feet, just as it is difficult to imagine top-flight marathon runners heaving 16-pound lead balls 60 feet. The correlation in the bottom panel is near zero—0.02 to be exact—indicating no relationship between these variables. That, too, makes sense. As far we are aware, your heft has little to do with the reproductive capacities of your siblings.

So we have a handy number that can summarize some obvious relationships between body weight and just about any other variable you care to imagine. What exactly does this have to do with politics? Look back at that bottom panel of Figure 1.1. While it is hard to discern a clear relationship, one does exist. There is a correlation of +0.20 between conservatism and conscientiousness. What this means is that the higher your score on conservative issue positions, the higher your score on the conscientiousness index. This, though, is a pretty small correlation. A correlation of 0.20 means conscientiousness is positively related to conservatism, but there is a good deal of "slop"—in other words, plenty of conscientious people are liberal and plenty of not so conscientious people are conservatives. On the basis of the evidence, we cannot say that all conservatives are conscientious—we would need a correlation of 1.0 to do so, and we are clearly some distance from that territory. Technically, we cannot even say that conscientiousness causes conservatism or vice versa—conservatism and conscientiousness might both be caused by something else. What we *can* say is that there is a modest but systematic tendency for conscientious people to be conservative. That might not be completely obvious from the scatterplot in Figure 1.1, but it is there.

You might notice that underneath each figure reference is not just an r but also a p value. *P* here stands for probability and should be interpreted as the likelihood of a relationship occurring by chance. A low p increases confidence in a relationship. Scholarly norms hold that the p should be less than 0.05 (less than one chance in 20 that the relationship occurred by chance), as is the case for the conscientious-conservative connection, for the relationship to be considered meaningful or "statistically significant." So, to vastly oversimplify, in evaluating relationships, look for big r values and low p values.

A correlation coefficient of 0.20 may seem limp and anticlimactic but in the world of social behavior, coefficients of 0.20 or even 0.10 are often met with great excitement, especially when they are replicated by other researchers. For example, traits or behaviors that demonstrate statistically significant correlations with a serious health issue, say breast cancer, of even .1 are viewed as quite important. Ultimately, this is the reason we have taken a statistical digression in the first chapter of the book and run the risk of sending you fleeing back into the comforting polemics of Kenny Conservative or Linda Liberal. The vast majority of the relationships we are going to describe in this book can be summarized by similarly modest correlations. If you think you are an exception to one of the correlations reported in these pages, you are probably right and undoubtedly have a good deal of company. This does not mean those relationships are any less real, though, as long as you remember to think probabilistically. Just as one cold day does not falsify global warming, one conscientious liberal does not alter the fact that there is a systematic relationship between conservatism and conscientiousness.

Leibniz's Baloney: What Is a Predisposition?

Looking at a pencil is not exactly a thrill-a-minute proposition, but what happens inside your body during this mundane event is a slick piece of biochemical engineering. The eye treats the shape and color of the object as input that is transmitted via the optic nerve to the occipital lobe at the back of the brain where it is then relayed to other parts of the brain and identified as a pencil. What doesn't happen is also interesting. Though your neurobiology is involved, viewing the pencil likely does not stir up much activity in the limbic or emotional parts of your brain. Looking at pencils, in other words, does not typically give people joy, melancholy, or a case of the hots. The limbic system simply can't be bothered to get out of bed to pay a pencil much mind. People know this—if you show them a pencil and ask the degree of emotional intensity felt, people's typical answer is "very little." This lack of interest can also be measured biologically; physiological changes recorded while a pencil is being viewed tend to be minimal or nonexistent.

Other objects and concepts are not like pencils and stir the brain's emotional parts from slumber. Loved ones, dangerous animals, beautiful landscapes, disgusting objects, cute babies, and threatening situations all tend to spur activity in neural channels not activated by viewing a pencil. In response to such stimuli, people report intense reactions and exhibit physiological changes.

Brain imaging will show heightened activity in emotionally relevant parts of the brain, including the amygdala, insular cortex, hypothalamus, hippocampus, and anterior cingulate cortex; an endocrinological assay will show alterations in hormonal levels; heart rate and respiration will accelerate, pupils will dilate, and palms will get sweaty. To put it more simply, the body changes in measurable ways.

These physiological changes affect how an object is perceived, processed, and responded to—and the variation from person to person in the nature of these responses is substantial. Each of us is primed to respond physiologically and psychologically to certain categories of stimuli—just not to the same stimuli and not to the same degree. Show a group of people the same stimulus and some will flatline while others will get a case of the vapors. These standing, biologically instantiated response orientations are a key part of what we mean when we say "predispositions." In sum, people are characterized by biological and psychological response tendencies, patterns that can be overridden but that serve as important shapers of behavior.

People are not fully conscious of their predispositions. Gottfried Leibniz, a seventeeth-century mathematician and scientist, called them "appetitions" and argued that, though unconscious, appetitions drive human actions. His ideas so troubled Descartes-addled Enlightenment minds that they were not published until well after Leibniz's death. Even then, they were not taken seriously for a long time. Recent science, though, is fully on board with Leibniz's ideas and is providing ever-increasing evidence that people grossly overestimate the role in their decisions of rational, conscious thought, just as they grossly overestimate the extent to which sensory input is objective.

Neuroscientist David Eagleman goes so far as to claim that "the brain is properly thought of as a mostly closed system that runs on its own internally generated activity ... internal data is not generated by external sensory data but merely modulated by it."[14] Noting that people often do things because of forces of which they are not aware and then produce a bogus reason for these actions after the fact, Stephen Pinker refers to the portion of the brain involved in constructing this post hoc narrative as the "baloney generator."[15] The baloney generator is so effective that people believe they know the reasons for their actions and beliefs even when these reasons are inaccurate and patently untrue.[16]

Need examples of physiology affecting attitudes and behavior, even when people think they are being rational? Consider this: Job applicant resumes reviewed on heavy clipboards are judged more worthy than identical resumes on lighter clipboards; holding a warm or hot drink can influence whether opinions of other people are positive or negative; when people reach out to pick up an orange while smelling strawberries they unwittingly spread their fingers less widely—as if they were picking up a strawberry rather than an orange.[17] People sitting in a messy, smelly room tend to make harsher moral judgments than those who are in a neutral room; disgusting ambient odors also increase expressed dislike of gay men.[18] Judges' sentencing practices are measurably more lenient when they are fresh and haven't just dealt with a string of prior cases.[19] Sitting on a hard, uncomfortable chair leads people to be less flexible in their stances than if they are seated on a soft, comfortable chair, and people reminded of physical cleansing, perhaps by being located near a hand sanitizer, are more likely to render stern judgments than those who were not given such a reminder.[20] People even can be made to change their moral judgments as a result of hypnotic suggestion.[21]

In all these cases the baloney generator can produce a convincing case that the pertinent decision was made on the merits rather than as a result of irrelevant factors. People actively deny that a chunky clipboard has anything to do with their assessment of job applicants or that a funky odor has anything to do with their moral judgments. Judges certainly refuse to believe that the length of time since their last break has anything to do with their sentencing decisions; after all, they are meting out objective justice. Leibniz was right, though, and the baloney generator is full of it. The way we respond—biologically, physiologically, and in many cases unwittingly—to our environments influences attitudes and behavior. People much prefer to believe, however, that their decisions and opinions are rational rather than rationalized.

This desire to believe we are rational is certainly in effect when it comes to politics, where an unwillingness to acknowledge the role of extraneous forces of which we may not even be aware is especially strong. Many pretend that politics is a product of citizens taking their civic obligations seriously, sifting through political messages and information, and then carefully and deliberately considering the candidates and issue positions before making a consciously informed decision. Doubtful. In truth, people's political judgments are affected by all kinds of factors they assume to be wholly irrelevant.

Compared to people (not just judges) with full stomachs, those who have not eaten for several hours are more sympathetic to the plight of welfare recipients.[22] Americans whose polling place happens to be a church are more likely to vote for right-of-center candidates and ideas than those whose polling place is a public school.[23] People are more likely to accept the realities of global warming if their air conditioning is broken.[24] Italians insisting they were neutral in the lead-up to a referendum on expanding a U.S. military base, but who implicitly associated pictures of the base with negative terms, were more likely to vote against the referendum; in other words, people who genuinely believed themselves to be undecided were not.[25] People shown a cartoon happy face for just a few milliseconds (too quick to register consciously) list fewer arguments against immigration than those individuals who were shown a frowning cartoon face.[26] Political views are influenced not only by forces believed to be irrelevant but by forces that have not entered into conscious awareness. People think they know the reasons they vote for the candidates they do or espouse particular political positions or beliefs, but there is at least a slice of baloney in that thinking.

Responses to political stimuli are animated by emotional and not always conscious bodily processes. Political scientist Milt Lodge studies "hot cognition" or "automaticity." His research shows that people tag familiar objects and concepts with an emotional response and that political stimuli such as a picture of Sarah Palin or the word "Obamacare" are particularly likely to generate emotional or affective (and therefore physiologically detectable) responses. In fact, Lodge and his colleague Charles Taber claim that "all political leaders, groups, issues, symbols, and ideas previously thought about and evaluated in the past become affectively charged—positively or negatively."[27] Responses to a range of individual concepts and objects frequently become integrated in a network that can be thought of as the tangible manifestation of a broader political ideology.

The fact that extraneous forces that may not have crossed the threshold of awareness (sometimes called sub-threshold) shape political orientations and actions makes it possible for individual variation in nonpolitical variables to affect politics. If hotter ambient temperatures in a room increase acceptance of global warming, maybe people whose internal thermostats incline them to feeling hot are also more likely to be accepting of global warming. Likewise, sensitivity to clutter and disorder, to smell, to disgust, and to threats becomes potentially relevant to political views. Since elements of these sensitivities often are outside of conscious awareness, it becomes possible that political views are shaped by psychological and physiological patterns.

It is important to recognize that predispositions are not fixed at birth. We cannot emphasize enough that we are *not* making a nature versus nurture argument. Innate forces combine with early development and later powerful environmental events to create attitudinal and behavioral tendencies. These predispositions are physically grounded in the circuitry of the nervous system, so once instantiated they can be very difficult, but far from impossible, to change. Altering a predisposition is like turning a supertanker; it usually takes concerted force for an extended period of time, but it can be done. Just like those heavy clipboards, a variety of predispositions nudge us in one direction or another, often without our knowledge, increasing the odds that we will behave in a certain way but leaving plenty of room for predispositions to be contravened and also for the predispositions themselves to be modified.

Still, while it is possible for situations and events to alter predispositions, attitudes are notoriously resistant to change. This is true outside the realm of politics and definitely true within it.[28] An individual's political orientation follows a pattern similar to that identified for happiness. Psychologists frequently refer to a "happiness set point." Events throughout a lifetime make people happier or sadder for a time but most individuals are generally oriented toward being upbeat or not and the effects of various events typically lead to modest and temporary deviations from the set point. Several months after experiencing even major life events such as an amputation or winning the lottery it appears that most people have returned to a degree of happiness with their lives surprisingly similar to that present before the major event.[29]

Politically relevant predispositions are similar: malleable but also resistant to change. This conclusion squares with a growing body of evidence documenting the long-term stability of people's political orientations.[30] Most people know someone who has done a political 180-degree turn, but these individuals stand out because they are relatively rare and do not pose a challenge to the core idea of predispositions as physically instantiated inclinations (remember, think probabilistically). We believe the reason for this relative stability is the existence of an ingrained emotional and therefore physiological response to stimuli that ends up being relevant to politics. It takes quite a bit for such habituated emotional responses to be eliminated, let alone reversed. Once they are there, they tend to be there for the long haul. As one study concludes, "[W]hen it comes to politics you've either got it or you don't."[31]

Predispositions, then, can be thought of as biologically and psychologically instantiated defaults that, absent new information or conscious overriding, govern response to given stimuli. For example, people may have a predisposed response to Barack Obama that would be evoked by a garden-variety image of him. Subsequent events and information, perhaps about his role in killing Osama Bin Laden, or a picture of him losing his composure, could alter that default response. The question is whether the new information becomes a long-term component of (adjusted) predispositions or whether, say, an existing negative predisposition toward Barack Obama would soon neutralize the positivity that might have been generated by the successful attack on Bin Laden's compound, rendering the new information irrelevant to evaluations.

A final critical and often misunderstood element of predispositions is that they are not equally present in all people. Just as the content of the predisposition varies from person to person, so too does the degree to which a predisposition is present at all. Being politically predisposed is not a requirement for membership in the human race. Like most everything else, the presence of predispositions should be thought of as operating along a continuum. Certain people are in possession of powerful political predispositions and politically relevant stimuli set off easily measurable psychological, cognitive, and physiological responses. Perhaps the nature of the political predisposition points in a liberal direction, perhaps in a conservative direction, or perhaps in different directions depending upon the particular issue. Other people have much weaker political predispositions. For them, politics is mostly irrelevant and they do not have much in the way of a preexisting, physiocognitive basis for their political behavior and attitudes. These individuals are often puzzled by all the fuss about politics.

The central thesis of this book is that many people have broad predispositions relevant to their behaviors and inclinations in the realm of politics. These predispositions can be measured with psychologically oriented survey items, with cognitive tests that do not rely on self-reports, with brain imaging, or with traditional physiological and endocrinological indicators. Due to perceptual, psychological, processing, and physiological differences, liberals and conservatives, for all intents and purposes, perceive and thus experience different worlds. Given this, it is not surprising to find they approach politics as though they were somewhat distinct species.

"We Have Known That All the Time!" But "It Can't Be True!"

These claims create controversy inside and outside academia but also seem intuitive. Folk wisdom has long put down political differences to something deep, perhaps even biological. Groucho Marx famously remarked that "all people are born alike—except Republicans and Democrats." In their comic opera "Iolanthe," Gilbert and Sullivan wrote that "every boy and every gal that's born into the world alive, is either a little liberal or else a little conservative." Enduring political differences are endless grist to the mill of humorous one-liners: Democrats eat their fish, Republicans hang theirs on the wall; Democrats make plans and do something else, Republicans follow the plans their grandparents made; Republicans tend to keep their shades drawn although there is seldom any reason why they should, Democrats ought to but don't.[32]

Folk wisdom may recognize the deep, nonpolitical differences separating liberals from conservatives but academic wisdom is not so sure. There have been numerous efforts to study whether political beliefs reflect deeper psychological tendencies such as personality traits (we address this possibility in Chapter 4). These attempts have frequently been met with scorched earth criticism. In the 1950s Theodor Adorno's book on the authoritarian personality was derided as "the most deeply flawed work of prominence in political psychology," the "Edsel of social science research," and one of the most harmful books in centuries.[33] In the 1960s, Silvan Tomkins' theories of biologically based emotions and their potential links to political temperament were subject to vigorous counterattacks.[34] In the 1960s and 1970s when Glenn Wilson and John Patterson developed a general instrument of social conservatism and argued it represented an underlying personality trait that was possibly heritable,[35] they were immediately challenged.[36] Comedians, songwriters, and the lay public have long taken for granted that politics runs deep and connects to other facets of life, but historically many in the academic community have been unwilling to concede this point.

This situation may finally be changing. After a lull, the last 10 years have seen a flowering of research on the broader forces intertwined with politics. This more recent research can be placed in one of two overarching categories. In the first, politics is measured using survey questionnaires that probe characteristics like personal values, moral foundations, personality traits, psychological tendencies, and sensitivity to disgust.[37] In the second, students of political orientations employ a whole array of cognitive and biological tests, including eyetracking, gaze cuing, brain imaging, genetics, electrodermal activity, and electromyography (facial muscle movements). These techniques make it possible to acknowledge the important role of factors that may not enter people's conscious awareness.

Survey-type self-reports are important parts of the measurement arsenal but sometimes people's baloney generators get in the way and leave them incapable of reporting how they feel and why they did what they did. The predispositions people bring with them into political situations can be referred to as motivated social reasoning, hot cognition, habits, longstanding predispositions, or antecedent conditions.[38] Regardless of the label, the nature of these predispositions is in part unavailable to the people holding them; as a result, techniques other than survey self-reports are essential and form a central element of our story. Even readers primed to accept that the differences between liberals and conservatives extend well beyond the realm of politics may not appreciate the biological and cognitive depth of these differences.

In short, for the first time real progress is being made in connecting political variations with biological and cognitive variations. This newer, biologically informed research is cumulating in a fashion that the more psychologically based efforts of the 1950s, 1960s, and 1970s did not, but the critics of placing politics in broader context have not gone away. In fact, for several reasons, they appear more tenacious. The incorporation of biology is particularly troubling to people who fail to realize that biology is not tantamount to determinism. Many scholars believe the only way to understand political orientations is "by understanding history and culture." They believe the notion that biologically instantiated predispositions have a universal application to politics regardless of time and space is "implausible" or even "incoherent."[39] We assert instead that history and culture have helped to shape biologically instantiated predispositions that then take on a life of their own and need to be studied alongside history and culture.

Moreover, the popular press monitors academic findings in this area closely, opening channels to a broader array of critics. Online outlets and networks further widen opportunities to offer commentary, particularly on an incendiary topic such as the deeper differences between liberals and conservatives. Whereas critics of the earlier iterations often were restricted to academic circles, that is hardly an apt description of the current situation. George Will assailed psychologist John Jost for his assertion that political orientations are undergirded by motivated social cognition. Will apparently took umbrage at the notion that people are merely led around by their "dispositions."[40]

For their part, bloggers claim that political choices are made "according to what is good and evil" and they often challenge the evidence documenting the importance of predispositions. They insist that political beliefs are fully under an individual's control, meaning that people who hold "terrible" beliefs can be expected to "come around eventually," though such a belief is more wishful thinking than factual. Whether the worry is that the existence of deeper, biological, politically relevant predispositions will impugn people's preferred political ideology or, more generally, that it will call into question the ability of humans to be politically rational and decent, denial of the existence of politically relevant predispositions is common.

Though critics of the movement to place politics in biological and psychological context hail from academia, journalism, and the public at large, political scientists are especially dubious. A longstanding assumption in political science, best exemplified in the influential work of Philip Converse, is that political beliefs and ideologies are narrow and apply only to politics. The fundamental idea is that to be in possession of a political ideology it is necessary to know the meaning of labels such as liberal and conservative and also to be in possession of a consistent set of political preferences that add up to a coherent match with those labels. The notion that people's politics bubble up from their broader, inner machinery is absent from the Converse view and therefore from much of traditional political science. As a result of this formulation, many scholars have convinced themselves that ideology is rare and getting rarer now that the big isms, such as Communism and Fascism, are history.[41] The gist is that, at least politically, people now are all the same, residing in a motley middle where they are undisturbed by the flow of big ideas and very disturbed by the ideological bent of elite politicians.[42] Only the elites have ideologies, we are told, and claims that politics is an extension of each individual's unique and generic psychological, cognitive, and physiological forces are treated with a heavy dose of skepticism (which we don't mind) and even derision and contempt (which we do).

In sum, Groucho Marx, Gilbert and Sullivan, and folk wisdom notwith-standing, plenty of people find the possibility of deeper, biological bases of politics both unbelievable and off-putting. Journalist and author Chris Mooney captures the situation when he describes the assertion that liberals and conservatives are different sorts of people as "something we've always sort of known but never really been willing to admit."[43] Our own research has been dismissed as inconsistent with realistic beliefs about humans and politics and simultaneously written off as something that "we already know," even though we're not sure how it's possible to be guilty of both sins at once. Our goal in this book is to show readers that deep, biological, politically relevant predispositions are quite real and anything but preposterous.

Conclusion: Why Can't We All Just Get Along?

Though seductive, the vision of a political rapprochement in which individuals from all corners of the polity converge in the political middle as they sing "Kumbaya" is a dangerous fantasy. Even if such a group-sing came to pass, liberals probably would be holding their lighters aloft, swaying as they sang with undisciplined abandon, while conservatives would be sitting in orderly rows, perhaps pews, performing a clipped, somewhat cold, but extremely well-rehearsed rendition. The forced agreement on lyrics and melody would be superficial and misleading. Vidal would be in the back making up dirty lyrics and Buckley would be down front trying to maintain order and threatening to punch Vidal in the kisser. The way to live with political differences is not to perpetuate the myth that they are a passing and remediable inconvenience but to recognize their depth and work effectively within the constraints they inevitably create. Rather than fanning the flames of ideological disagreement, the goal should be to ameliorate the problems disagreement creates.

Acknowledging the real nature of people's political differences could allow strategies for campaigns and for the presentation of governmental policies to be more fine tuned, targeted, and effective; policies themselves could be more legitimate as a result, thereby facilitating compliance; casual political discourse could also become more constructive and, perhaps most importantly, understanding and tolerance of those who disagree with us politically could be enhanced, rendering the entire political arena a less frustrating place. Understanding the reasons for gridlock and polarization will not cause these problems to disappear magically but will suggest realistic approaches to softening their edges and improving governance. In Chapter 9, we will describe in more detail how all this might work.

Such an acknowledgment would not entail giving up on efforts at political persuasion. Remember that the relationships we are about to describe are modest and probabilistic. Large numbers of people do not have clear predispositions toward the political right or left and these people are "in play." Efforts at persuasion should continue even though those with politically relevant predispositions will be difficult to turn. Approval of the other side is not what we advocate but the political system will be a happier and more productive place if political adversaries are viewed not with scorn but with a perhaps grudging recognition that they experience a different world.

This means accepting that political orientations are connected to deep physiocognitive predispositions in a broadly predictable fashion. Acceptance of this belief requires rejecting two widely accepted assertions. The first is that all politics is culturally and historically idiosyncratic since one society might be concerned with famines and droughts, another with the superpower across the river, and yet another with protecting mineral riches. If this assertion is true, it becomes pointless to try to generalize about political divisions, patterns, and viewpoints. The second assertion is that, though humans' physical traits obviously vary, we all share the same basic psychological, emotional, and cognitive architecture. If, from a behavioral point of view, human architecture is all the same, it follows that differences in political orientations cannot be more than skin deep and physiocognitive predispositions are irrelevant.

Both assertions—one about the nature of politics and one about the nature of humans—are incorrect. In fact, they have it exactly backwards: Though traditional wisdom asserts that politics varies and human nature is universal, in truth politics is universal and human nature varies. Failing to appreciate these two points renders it impossible to grasp the true source of political conflict. Accordingly, before we present empirical evidence documenting the deep-seated psychological, cognitive, physiological, and genetic correlates of political variation (Chapters 4–7), we first need to make the case that politics is universal and human nature is variable (Chapters 2 and 3, respectively).

Notes

[1] You can find a complete audio of the 22-minute debate at http://www.pitt.edu/~kloman/debates.html. Clips of the juiciest exchanges can be found on Youtube.

[2] Miller, *The Wolf by the Ears*, 148–151.

[3] Johnson, *The Information Diet*.

[4] O'Reilly, *Pinheads and Patriots: Where You Stand in the Age of Obama*; and Franken, *Rush Limbaugh Is a Big Fat Idiot*.

[5] Haston, *So You Married a Conservative: A Stone Age Explanation of Our Differences, a New Path towards Progress*, 3.

[6] McCloskey, "Conservatism and Personality," 37.

[7] Block and Block, "Nursery School Personality and Political Orientation Two Decades Later."

[8] Munro, "The Scientific Impotence Excuse."

[9] Mill, *On Liberty*.

[10] Emerson, "The Conservative."

[11] Bobbio, *Left and Right*; and Jost, "The End of the End of Ideology."

[12] Lane, *Political Ideology: Why the American Common Man Believes What He Does*.

[13] We are massively oversimplifying—validating psychometric scales is a big industry in social science that employs scary levels of statistical sophistication. Still, asking carefully vetted questions and adding them up is the basic gist of it.

[14] Eagleman, *Incognito: The Secret Lives of the Brain*, 44.

[15] Pinker, *The Blank Slate: The Modern Denial of Human Nature*.

[16] Gazzaniga, *Who's In Charge?: Free Will and the Science of the Brain*.

[17] Castiello et al., "Cross-Modal Interactions between Olfaction and Vision When Grasping"; and Ackerman et al., "Incidental Haptic Sensations Influence Social Judgments and Decisions."

[18] Schnall et al., "With a Clean Conscience: Cleanliness Reduces the Severity of Moral Judgments"; Inbar et al., "Conservatives Are More Easily Disgusted Than Liberals"; and Inbar et al., "Disgusting Smells Cause Decreased Liking of Gay Men."

[19] Danziger et al., "Extraneous Factors in Judicial Decisions."

[20] Ackerman et al., "Incidental Haptic Sensations Influence Social Judgments and Decisions"; and Helzer and Pizarro, "Dirty Liberals! Reminders of Physical Cleanliness Influence Moral and Political Attitudes."

[21] Wheatley and Haidt, "Hypnotic Disgust Makes Moral Judgments More Severe."

[22] Michael Bang Petersen, personal communication.

[23] Berger et al., "Contextual Priming: Where People Vote Affects How They Vote"; and Rutchick, "Deus Ex Machina: The Influence of Polling Place on Voting Behavior."

24 Risen and Critcher, "Visceral Fit: While in a Visceral State, Associated States of the World Seem More Likely."

25 Galdi et al., "Automatic Mental Associations Predict Future Choices of Undecided Decision-Makers."

26 Milt Lodge, personal communication.

27 Lodge and Taber, "The Automaticity of Affect for Political Leaders, Groups, and Issues: An Experimental Test of the Hot Cognition Hypothesis," 456.

28 Ditto and Lopez, "Motivated Skepticism: Use of Differential Decision Criteria for Preferred and Nonpreferred Conclusions"; Edwards and Smith, "A Disconfirmation Bias in the Evaluation of Arguments"; Munro et al., "Biased Assimilation of Sociopolitical Arguments: Evaluating the 1996 U.S. Presidential Debate"; Zaller, *The Nature and Origins of Mass Opinion*; Marcus et al., *With Malice toward Some: How People Make Civil Liberties Judgments*; and Gerber et al., "Voting May Be Habit-Forming: Evidence from a Randomized Field Experiment."

29 Fujita and Diener, "Life Satisfaction Set Point: Stability and Change."

30 Gerber et al., "Voting May Be Habit-Forming: Evidence from a Randomized Field Experiment"; and Sears and Funk, *The Role of Self-Interest in Social and Political Attitudes*, in *Advances in Experimental Social Psychology*.

31 Prior, "You've Either Got It or You Don't? The Stability of Political Interest over the Life Cycle."

32 More systematically, when political scientist Leonie Huddy (personal communication) asked a random sample of survey respondents to describe the difference between liberals and conservatives, the most common difference people mentioned was not political views but personality traits. See also Conover and Feldman, "The Origins and Meaning of Liberal/Conservative Self-Identifications."

33 Martin, "The Authoritarian Personality, 50 Years Later: What Lessons Are There for Political Psychology?"; Roisier and Willig, "The Strange Death of the Authoritarian Personality: 50 Years of Psychological and Political Debate"; and Wolfe, "The Authoritarian Personality Revisited."

34 Kosofky et al., "Shame in the Cybernetic Fold: Reading Silvan Tomkins."

35 Adorno et al., *The Authoritarian Personality*; Tomkins, "Left and Right: A Basic Dimension of Ideology and Personality"; and Wilson, *The Psychology of Conservatism*.

36 Ray, "How Good Is the Wilson and Patterson Conservatism Scale?"

37 Caprara et al., "Personality Profiles and Political Parties"; Jost et al., "Political Conservatism as Motivated Social Cognition"; Chirumbolo et al., "Need for Cognitive Closure and Politics: Voting, Political Attitudes and Attributional Style"; Graham et al., "Liberals and Conservatives Rely on Different Sets of Moral Foundations"; Inbar et al., "Conservatives Are More Easily Disgusted Than Liberals"; Golec et al., "Political Conservatism, Need for Cognitive Closure, and Intergroup Hostility"; Schwartz et al., "Basic Personal Values, Core Political Values, and Voting: A Longitudinal Analysis"; Mondak, *Personality and the Foundations of Political Behavior*; and Haidt, *The Righteous Mind*.

38 Zajonc, "Feeling and Thinking: Preferences Need No Inferences"; Lodge and Hamill, "A Partisan Schema for Political Information Processing"; Zaller, *The Nature and Origins of Mass Opinion;* Marcus et al., *With Malice toward Some: How People Make Civil Liberties Judgments*; Plutzer, "Becoming a Habitual Voter: Inertia, Resources, and Growth in Young Adulthood"; Gerber et al., "Voting May Be Habit-Forming: Evidence from a Randomized Field Experiment"; Lodge and Taber, "The Automaticity of Affect for Political Leaders, Groups, and

Issues: An Experimental Test of the Hot Cognition Hypothesis"; and Jost, "The End of the End of Ideology."

39   Charney, "Genes and Ideologies," 300.

40   Summarized in Jost, "The End of the End of Ideology."

41   Shils, "Authoritarianism: Right and Left," in *Studies in the Scope and Method of "The Authoritarian Personality"*; Bell, *The End of Ideology: On the Exhaustion of Political Ideas in the Fifties*; Converse, "The Nature of Belief Systems in Mass Publics"; and Fukuyama, *The End of History and the Last Man*.

42   Converse, "The Nature of Belief Systems in Mass Publics"; and Fiorina et al., *Culture War? The Myth of a Polarized America*.

43   Mooney, *The Republican Brain: The Science of Why They Deny Science—and Reality*.

CHAPTER 2

Getting Into Bedrock with Politics

If the Left-Right distinction did not exist, scholars of ideology would need to invent its equivalent.

John Jost

Politics is for the present . . . an equation is something for eternity.

Albert Einstein

Former U.S. Senator and candidate for the 2012 Republican presidential nomination Rick Santorum once described his country's universities and colleges as "indoctrination mills" for godless liberalism.[1] These strong words reflect the widespread suspicion among conservatives—and not just conservatives in America—that universities are less focused on raising IQs than they are on raising left-leaning consciousness. As long-time college professors, we are dubious. Persuading students to stop updating their Facebook pages long enough to listen to a 55-minute lecture is challenge enough; persuading large portions of them to pledge undying fealty to a particular political belief system strikes us as a fool's errand.

Still, this does not mean that conservative suspicions about faculty politics are without merit (most academics *are* left-leaning) or that there are no historical examples of campus ideological indoctrination. The City College of New York in the mid-twentieth century, for instance, came about as close as any institution of higher education will ever come to fulfilling right-wing night-mares of academia. The faculty, already tainted with a hint of radical leftism, caused a scandal by trying to hire British polymath Bertrand Russell—who apart from being a genius was a well-known Socialist, pacifist, and general promoter of avant-garde social ideas (he thought religion outdated and saw nothing morally objectionable about premarital sex). Scandalized citizens worried about Russell spreading his dangerous notions amongst New York's vulnerable youth and sued to prevent his hiring. Astonishingly, the legal system obliged. State Supreme Court Justice John McGeehan ruled Russell morally unfit to teach, the upshot being that City College students dodged the bullet of taking instruction from a future Nobel laureate.[2]

While successful at keeping Russell out, neither jurists nor citizens could prevent students from attending City College. This was unfortunate for champions of conservative rectitude in higher education; the students, if anything, were more radical than the faculty. Communists controlled the school newspaper; Socialists sought the ouster of the Reserve Officer Training Corps; and campus left-wingers of various denominations issued manifestoes denouncing capitalism, cuts in education, oppression of the working class, imperialist wars and nonimperialist wars, imperialists in general and Franklin Delano Roosevelt in particular, who apparently was considered by a surprising fraction of the student body to be an imperialist, right-wing, war-mongering oppressor of the working classes who was not doing nearly enough for education.[3]

Ground zero for all this hard left-wing activism was the City College lunchroom, where radicals and political activists of various stripes gathered to debate the finer points of Marxism, Socialism, Communism, and Trotskyism, not to mention Marlenism and Fieldism. The atmosphere and denizens of the lunchroom are fondly recalled in a semi-famous 1977 *New York Times Magazine* essay entitled, "Memoirs of a Trotskyist."[4] Apparently, it was a rundown place from an aesthetic point of view and was filled with lower- to middle-class Jewish students, mostly sons of immigrants who had brought their left-wing politics from Europe. At the time, anti-Semitism led to Jewish quotas at many American universities but not at liberal-minded City. As a result of the prejudices elsewhere in higher education, City College ended up with an astonishing concentration of intellectual talent, including nine Nobel Prize winners who graduated between 1935 and 1954.

At the periphery of the lunchroom were alcoves consisting of benches facing low refectory tables in rectangular or semicircular spaces. There were a dozen or so of these alcoves and each was the turf of a particular political, ethnic, or religious group; for example, there was a Zionist alcove, a Catholic alcove, and an alcove for the smattering of African American students. The biggest "political" alcove was Alcove No. 2, home turf of the Stalinists. These were mostly hardcore supporters of the type of Communism practiced by the Soviet Union. Alcove No. 2 regulars glorified Joseph Stalin and apparently spent a good deal of their time torturing facts and logic into supporting their preferred portrait of Uncle Joe as a benevolent and wise protector of the proletariat. Alcove No. 1, just to the right as you entered the cafeteria, was also a political alcove and also populated by leftists. These leftists, though, did not impose the same sort of ideological purity test required for admission into Alcove No. 2. They included a group of a dozen Trotskyists, a roughly equal number of Socialists, a few followers of other miscellaneous isms, and a handful of right-wingers, which in this group meant they voted for Roosevelt and called themselves Social Democrats. Radical left-wing politics and ideology was constantly discussed and debated in Alcoves No. 1 and No. 2, and the students doing the debating took their arguments out of the lunch room, periodically mounting protest rallies, and carrying their interpretations of Marx, Engels, Lenin, and Trotsky into classes taught by low-paid, liberal-leaning faculty.

If you believe conservative worries about higher education's impact on political beliefs, then surely you would expect students marinating in City College's left-wing stew for four years to infect the body politic with their "godless liberalism." You could even produce some evidence to support this belief. Julius Rosenberg, Communist boogeyman number one of the McCarthy era, was executed in 1953 for passing on atomic secrets to the Soviet Union. Before trying to advance the vanguard of the proletariat by giving Commies the bomb, Rosenberg had graduated from City College with a degree in electrical engineering. More principled and moderate leftists who were City College alums included people like Irving Howe, who went on to help found the quarterly magazine *Dissent* as well as the Democratic Socialists of America. Still, Rosenberg's lasting impact on politics was pretty much nil and Howe, for all his brilliance as a cultural critic, never managed to kick start a movement with any broad or lasting impact on politics.

That is not to say a movement failed to materialize from the radicalized, left-wing atmosphere of City College. A powerful and influential political movement was birthed, not in Alcove No. 2 but in Alcove No. 1, and not on the left but on the right. Alcove No. 1's most lasting political influence was what came to be known as the neoconservative movement. As such, its alumni and heirs influenced the politics of a generation, reshaped the policy orientations of a major American political party, and played an outsized role in promoting the interventionist foreign policies promulgated by the U.S. government during the very early portions of the twenty-first century, thereby molding American politics and radically altering other countries, from the USA Patriot Act to the war in Iraq. You see, a key player in Alcove No. 1 was Irving Kristol, described by *The Daily Telegraph* as "perhaps the most consequential public intellectual of the latter half of the 20th Century." So great was his influence on politics that one U.S. president joked that anyone seeking employment at the White House should just show up and say "Irving sent me."[5] That president was Ronald Reagan.

At least two lessons seem to flow from the political legacy of the radicals of Alcove No. 1. First, institutions of higher education cannot indoctrinate leftist political beliefs for toffee, even at a radicalized, left-leaning place like mid-twentieth-century City College. Several City alums who flirted with the politics of the radical left as students ended up all over the political spectrum as they got older and, it is fair to say, their most lasting political influence was not in promoting a Communist ideology but in promoting the right's "we are doing God's will" nationalism. And regardless of whether they kept to the left like Howe or drifted rightward like Kristol, their naviga- tion of the political spectrum was not put on automatic pilot by their experience as undergraduates.

The second lesson seems even clearer: Politics and political beliefs are fun- gible. They change depending on time and place. The Stalinist-Trotskyist split did not just demark who was welcome into Alcove No. 1 or No. 2; it held a central, vehement, and often violent place in the global politics of the hard left for much of the first half of the twentieth century. Nowadays? Well, not so much. It is difficult to find a true dyed-in-the-wool Marxist or Trotskyist evangelizing ideology on an American college campus these days. Those who do exist represent amusing or irritating relics of the past rather than existen- tial threats to the republic. Trotsky survives in college students' consciousness mostly in the names of punk rock bands. Moreover, an individual's preferences can evolve over time. Many giants of neoconservatism started out as liberals who supported the Democratic Party. They ended up as conservatives in the high echelons of the Republican Party.

We generally accept the first lesson: Colleges and universities stink at ideological indoctrination. There are enough counter-examples to keep an ember of righteous indignation glowing in certain circles, but you have to look pretty hard to find anyone doing this sort of thing with even moderate levels of success. Those who are any good at it are as likely to be on the right as the left; the academic neocons, for example, turned out to be a pretty persuasive bunch. We take issue, however, with the second lesson and contend that politics at its core is more or less invariable. As laid out at the end of the previous chapter, we are staking a claim that human nature is varied and politics is constant. Yet, how can politics be viewed as stable if the radical left can morph into the radical(ish) right, if the issues innervating students in the 1930s were not the issues innervating students in the 1960s or in 2010s? Gay marriage and global warming, for example, did not trip many triggers in the 1960s. Moreover, the political issues central to American political life at any given time are very different from the issues that occupy other nations and even the United States's closest allies. Abortion is an issue that polarizes politics in the United States but makes hardly a ripple in the United Kingdom. Tribal loyalties structure political orientations in some African nations but not n Denmark.

In this chapter we make the case that, despite the apparent idiosyncrasies of politics in the alcoves of City College and beyond, clear political commonalities are present everywhere and at all times if you look for them and are not philosophically opposed to finding them. These commonalities reside in what we refer to as the bedrock dilemmas of politics, and in order to understand these bedrock dilemmas it is helpful to go back in time a few thousand years.

I Am in Love with Your Political Orientation

If you can believe Diogenes Laertius, that gossipy ancient biographer of even more ancient philosophers, Aristotle was an odd-looking fellow with some even odder habits. He had thin legs, a lisp and small eyes. He was a flashy dresser, collected dishes, and was rumored to like taking baths in warm oil.[6] Aristotle, of course, is known less for his physical attributes and kinky hygiene habits than his considerable intellectual legacy, which has influenced everything from biology to philosophy. Political scientists are quick to claim dead white guys in togas as disciplinary forebears, and rather than give Aristotle his due as the first biologist, we like to view him as one of our own in part because of a famous line from early in *Politics*: "[M]an is by nature a political animal."[7]

A more precise translation of this aphorism is actually something like, "[M]an is by nature an animal intended to live in a polis," a polis being a city-state such as Athens or Sparta.[8] Exactly what Aristotle was getting at when he wrote those words is open to debate (*Politics* is a notoriously hard read). Still, it is reasonably clear that Aristotle was suggesting that man was a political animal at least in the sense that it is in man's nature to thrive in a mass-scale society. Though widely quoted as evidence that politics constitute a natural human activity, Aristotle actually took some pains to point out that a polis itself was not a wholly natural form of social organization. Aristotle believed that the polis—or what we would call a polity—was something new and different that developed from earlier forms of social organization—families, bands, and tribes of Yooks and Zooks. It was meta-social group, an association of associations.[9]

Though not a form of social organization found in nature, Aristotle was clearly onto something when he observed that humans take to living in a polity like ducks to water. There is something fundamental about politics, and it is deeply embedded in our nature. On this point we do not have to take the word of a spindly legged ancient soaking in a tub of olive oil. An increasing number of studies suggest that political beliefs and behaviors are so deeply embedded that they may be genetically influenced, and we devote an entire chapter to this research later on. Yet it is not just that political attitudes appear to be influenced by genes (many traits are); it is that genetics appear to exert more influence on political attitudes than other social attitudes. For example, genome-wide association studies (GWAS, pronounced as "gee-wahz") identify many places where the human genetic code varies—often a million or more—and then determine whether any of them line up in any systematic way with whatever trait a researcher is interested in. GWAS studies have started to examine a variety of social attitudes, and one of the more interesting findings is that genes seem to explain more about political preferences than, say, economic preferences.[10] We don't want to get too carried away here; our political beliefs are most definitely not genetically predetermined (remember, think probabilistically). Still, as we will explain in-depth later on, it appears that genes do predispose people toward certain political orientations, perhaps more than they do orientations in other areas.

More evidence of the centrality of politics comes from its influence on mate selection. Perhaps you think you married your wife because opposites attract or that you moved in with your boyfriend solely because of his winning personality. Think again. Social scientists of various stripes have spent a good deal of time examining who tends to form mate pairs with whom in order to obey a biological imperative to have kids, get a mortgage, and buy a minivan. What is crystal clear from this research is that people do not generally pair off with those who are similar to them in terms of personality traits—good news for us scholarly introverts. Some matching occurs on the basis of shared physical traits (height, weight, attractiveness) but even here the correlations are weak. If not personality and physicality, on what variables are mate pairs the most likely to be matches?

Easy. The top three variables on the "what traits do mate pairs most match up on" list are drinking, religion, and . . . drum roll . . . politics. (Education level is fourth.) What an interesting if explosive combination! Extroverts are as likely to marry introverts as they are to marry other extroverts but, James Carville and Mary Matalin aside, liberals are much more likely to marry other liberals than conservatives. Spouses tend to have similar political attitudes. But are mate pairs initially attracted to each other in part because of politics and values (what scholars call political assortative mating), or might there be other explanations for the fact that mates tend to be cut from the same political cloth? Two possible alternative explanations come to mind. The first is that when mate pairs begin their relationships they are not politically similar, but over the years and decades the political views of each mate "assimilate" to the political views of the other. The second possibility is that the "pool" from which an individual identifies a mate is probably similar in terms of sociodemographic traits (income, education, religion, region, and age) to the traits of the individual doing the identifying, and this "social homogamy" might lead to political similarities without politics playing a meaningful independent role. As plausible as they are, these alternative explanations for the political similarity of mates do not withstand analysis.

A comparison of the political similarity of couples married a short time and those married a long time indicates virtually no increased political similarity for longer marriages. One of the few data sets that make it possible to track couples over time comes to the same conclusion. In fact, on some issues, like gender roles, disagreement within mate pairs tended to become greater with the passage of the years (we bet that pattern makes for some interesting dinner table conversations). Finally, an intriguing data set from a large online data service shows that when political views are presented as options for selecting "matches" they are consulted eagerly, and decisions on the individuals with whom subsequent messaging will and will not occur are heavily influenced by similarity of political opinions. In sum, the high degree of political similarity is present from the very beginning of relationships.[11]

The social homogamy explanation fares no better than the assimilation explanation. When we analyzed a data set of several thousand mate pairs, we were surprised to discover that mates are politically similar even after controlling for all the sociodemographic variables that might define an individual's mating pool. What this means is that even if the analysis is restricted to narrow groups, mate pairs tend to be politically similar. For example, couples consisting of middle-class, college-educated, Midwestern Roman Catholics in their thirties tend to be much more politically similar than would be expected given the (substantial) political variation among individuals in that particular sociodemographic group.[12] Similar results have been obtained in work done outside the United States.[13] Apparently, regardless of country and culture, the mating game attaches importance to compatibility on political temperament even after controlling for the tendency of mate pairs to come from similar social groups.

While the deep and fundamental nature of politics brings us together, it also splits us apart. Political beliefs lead to astonishing acts of collective action for the sole purpose of punishing people with different political beliefs. It turns out that the primal urge driving William Buckley's publicly announced desire to sock Gore Vidal in the kisser is illustrative of an extremely powerful motivator of human social behavior. In Aristotle's day, for example, a polis would periodically band together with another polis or two to beat the bejeezus out of the next polis over. We humans are really big on trying to exterminate, or at least seriously annoy, another group because its members have different ideas about politics. Now, it is one thing to have someone steal your ass and then go looking to go kick his; it is quite another to kill complete strangers just because they have a tea towel flapping on a stick that is a different color than the tea towel flapping on your stick. Endow the tea towels with symbolic political importance, though, and plenty of us seem to be willing to do pretty much exactly that.

Politics does not divide us only on the mass-scale, but also on a much more personal level. Politics and its running mate, religion, tend to get people worked up to about the same pitch—which, to say the least, is high. This is why politics and not the pros and cons of extroversion is a taboo subject at many social gatherings. We can get sideways with people we love over things that may not have any meaningful relevance to either of our lives. Uncle Crusty might not know any gay couples but that does not stop him from ruining Aunt Sally's family reunion by denouncing them, veins bulging, at full volume.

Politics is deep and fundamental to humans; it defines us as a species and is likely, quite literally, in our DNA. Accepting that we are political animals, though, does not necessarily mean that politics is universal and stable across time and space. As Aristotle pointed out, a polis is not a natural form of social organization. It is a cultural construct, and while all polities are political, the particular issues and ideologies that animate alliances and divide families often seem to have little in common. The political beliefs that separated Athens from Sparta, Alcove No. 1 from Alcove No. 2, and Aunt Sally from Uncle Crusty are clearly very different. What could possibly connect these different political beliefs over the eons and around the globe?

Differences Galore?

Answering this question requires appreciating the differences between labels and issues of the day on the one hand and bedrock principles on the other. Issues include how much (and whom) to tax; the legality of abortion; social welfare; environmental regulation; and whether to go to war, to the moon, or to the International Monetary Fund with hat in hand. A complex mass-scale society can produce a virtually limitless supply of issues. Labels are simply the vocabulary employed to describe the reasonably systematic orientations toward issues that float around a polity at a given time. Labels can refer to actual

organizations such as political parties, or to less tangible entities such as ideologies (particular sets of beliefs about government and society).

To the casual observer—indeed, to most professional observers—issues and
labels constitute pretty much the entire content of politics. Issues are what people
argue and disagree about at a given time and place, and labels distinguish the groups
contesting those issues or the broad philosophical bases of those issue positions. If
you think of politics in terms of issues and labels, it is difficult to see anything that
looks universal and stable about it because issues and labels change across time
and from country to country. True, issues can dominate politics in a particular
place for an extended period of time. The legality of slavery, for example, was an
all-consuming issue in the United States for nearly the first hundred years of the
republic's existence. Eventually, though, even this issue faded—and more typical
issues do not have this kind of staying power. In fact, in these days of the 24-hour
news cycle issues can rise and fall between breakfast and lunch. Anyone remember
what big bruising fiscal issue separated candidates Al Gore and George W. Bush in
the 2000 U.S. presidential contest? No? Get this—it was what to do with the federal
government's budget surplus, an issue that has since gone the way of, well, the federal government's budget surplus.

The labels that organize issue disagreements likewise seem to be historically, culturally, or geographically idiosyncratic. In many countries the word
"liberal" refers to individuals supporting policies best characterized as mildly
libertarian: limited governmental involvement in social as well as economic
issues. In the United States, though, "liberal" is associated with economic positions that are anything but libertarian. Even the concept of a "left" and "right"
as a means of universally organizing political preferences seems to be bound
by time and culture. The origin of the left-right political divide is mostly the
product of seating arrangements of the 1789 Estates-General in Revolutionary France. It may be that, if *ancien régime*–supporting aristos had happened
to park their silken-clad bottoms to the King's left, Stalin and Lenin would be
remembered as hard core right-wingers and Hitler as the par exemplar of the
deranged left. If the toffs were seated in the galleries, maybe we would have
an up-down divide rather than a left-right (which might be fun—imagine the
signs at City College protests: "Down with Uppism!").

It is because issues and labels are so variable across time and space that many
political scientists are skeptical about the whole idea of ideology, especially the

notion that systematic sets of political beliefs can be neatly ordered along a dimension with moderates in the middle between two extremes of left and right. Traditionally speaking, the political left has been associated with support for equality and tolerance of departures from tradition, while the right is more supportive of authority, hierarchy, and order.[14] As political scientists have routinely pointed out, exceptions abound. Communists can be a pretty authoritarian bunch, though they are traditionally placed on the left, and conservatives are often fierce defenders of individual liberties even though they are viewed as residing on the right. Some people seem to simultaneously hold beliefs associated with the left and the right. Libertarians, for example, tend to be left leaning on many social issues (gay marriage, abortion), but right leaning on economic issues (government regulations, taxes).

Political scientists who study issue attitudes have frequently come to the conclusion that political beliefs are multidimensional and that where you are sitting on any particular issue at any particular time is determined wholly by social and cultural forces. In short, much research argues that people are simply not very ideological; that their political beliefs do not systematically add up to a stable and meaningfully interpretable point on a left-right dimension. Only a few people, this argument goes, wander the world with some underlying stable philosophy of government that charts where we sit in the grand hall of political beliefs and attitudes. Rather than stable philosophical or psychological gyroscopes, individual political beliefs and attitudes are seen as more a mash-up of purely social and environmental influences ranging from family, friends, schooling, and peers to whether you recently received a pink slip, find the president attractive, served in the military, or just woke up feeling patriotic this morning.[15]

According to this story, the left-right dimension is as arbitrary as the seating arrangements of a long-ago French parliament. It is "spurious" and a "poor description of political attitudes for the overwhelming proportion of people everywhere," since any actual mapping of political belief systems does not provide a neat, linear, one-dimension arrangement but "a jumbled cluster of pyramids ... with the mass bases of the pyramids overlapping in such profusion that it would be impossible to decide where one pyramid ended and another began."[16]

If political beliefs and even the vocabulary used to describe political beliefs have basically just been made up in the past couple hundred years to deal with whatever issues or ideas are floating around a given culture at a given time, the universality of politics looks very elusive indeed. We might be political animals, but people seem to acquire unique outlooks depending upon their particular cultures and historical niches.

Commonality Reigns! Political Universals

So, to sum up the story so far, on the one hand beliefs and issues are all over the place and only really hang together in any sensible sort of fashion if you trap them in a particular place at a particular time. On the other hand, at least since Aristotle, it has been recognized that humans are political by nature. Even though they live in unnatural agglomerations called polities, they universally take to the social relations of these associations of associations. Like bees and ants, we seem to be designed for social living, but unlike bees and ants we are not just social but political in the sense that politics is contentious and emotionally charged, and promotes conflict as well as cooperation. Politics is such a fundamental part of our natures that political temperaments are at least partially heritable and mates are selected on the basis of politics (thus further shaping the political temperaments of offspring). Heritable political beliefs make no sense if politics is purely the product of our social environments.

The key to understanding this apparent contradiction is what we call bedrock social dilemmas. These reflect divisions in the underlying first principles of politics: core preferences about the organization, structure, and conduct of mass social life. All social units face the same need to resolve certain social dilemmas. They need to decide on leadership and decision-making arrangements, distribution of resources, and how best to secure protection from out-groups, as well as punish the misbehavior of in-group members and orient to traditional (as opposed to new) forms of social behavior. People clearly have different underlying preferences regarding these bedrock dilemmas. Some prefer more hierarchical decision-making while others prefer egalitarian arrangements; some believe in share and share alike while others believe in taking care of your own; some see out-groups as threats while others see them as potential sources of friendship and new knowledge.

These underlying bedrock dilemmas of politics are fundamental to human social life; they are never fully resolved, and disputes over the best solutions to these dilemmas constantly churn human societies. Small hunter-gatherer societies needed to figure out the appropriate way to share the spoils of the hunt; large, developed modern societies need to settle on the niceties of tax codes and social welfare structures. Small hunter-gatherer societies needed to figure out the best way to treat one of their own who committed a serious violation of an established social norm: Death? Banishment from the group? Leave it up to the victim of the offense? Public embarrassment? Forgiveness? Large developed modern societies need to settle on the finer points of the criminal code: Three strikes and you're out? Probation? Loss of freedom? Community service? Restitution? Rehabilitation? Capital punishment? The issues sometimes appear quite different, but the bedrock principles are exactly the same.

Labels and issues are just waves on the surface; they can be whipped up and blown every which way by the winds of history and culture. What they are all created from, though, is the same basic set of underlying currents. These dilemmas have been tacitly recognized as the basis of politics in mass-scale societies for at least two thousand years. In *Politics*, Aristotle tackles a wide-ranging set of preferences for the structure and organization of the polity; he specifically undertakes an analysis not just of Athenian but of Spartan, Cretan, and Carthaginian approaches to running a polity and notes big differences in preferences for the structure and organization of mass-scale social life. Athens was run through a direct democracy where citizens, or at least well-off men, voted directly on issues. Sparta was more authoritarian, with hereditary monarchs and an elected-for-life council. Differences also appeared in resource distribution, social structure, and expected and enforced social behavior, as well as in differing sets of institutions. Viewed from the perspective of bedrock political dilemma, Sparta is to conservative as Athens is to liberal.

Aristotle's basic analysis can be applied to mass-scale social organizations everywhere. Over two thousand years later, these same underlying issues animated the discussions in Alcoves No. 1 and No. 2 at City College; nearly brought Buckley and Vidal to blows on national television; and define most of the disagreements between conservatives, liberals, Republicans, Democrats, Social Democrats, the National Front, and insert-label-of-familiar-party-ideology-or-political-group-here. How should our mass-scale society make decisions? What rules should everyone follow? What should we do with people who do not follow the rules? Do we try new things or stick with tradition? *This* is what we mean by the universality of politics. Regardless of the issues and labels, the same set of dilemmas lurks underneath.

When we talk about predispositions in this book, we are referring to standing orientations that have a measurable biological (though not necessarily innate) signature—in other words, to political predispositions that are biologically instantiated. The labels for particular political predispositions might be different in different places at different times, and they might be applied to wildly varying sets of issues, but that does not mean they are not reflecting a standard set of positions on bedrock dilemmas. We might call people conservatives and liberals in one place and kumquats and rutabagas in another—or even the same person a Democrat in one era and a Republican in another—but in all societies there are those who favor sticking with traditional values and those who favor more experimental social arrangements, those who want strong leaders and those who want a more egalitarian approach, those who advocate engagement with out-groups and those who see out-groups as threats to be avoided or conquered, those who call for resource redistribution and those who do not.

To get an idea of how a set of predispositions toward these bedrock dilemmas can provide a constant anchor underneath shifting issues and labels, consider contemporary attitudes toward military intervention, an issue much fought over by the left and right in the United States. Generally speaking, it has been the right—Republicans—that has been more supportive of this sort of thing, while the long hairs on the left are more opposed. Military action in Iraq, Afghanistan, Panama, and Grenada was launched by Republican presidential administrations, and a big chunk of the public support for those operations was provided by rank-and-file Republicans. Vietnam was fought primarily under Democratic administrations, but the war was strongly supported by many Republicans. The domino theory that Communism, if not stopped in Saigon, would spread to the rest of Southeast Asia, then Honolulu, and then Berkeley ("too late," many Republicans would argue) was widely advocated by the 1964 Republican presidential candidate Barry Goldwater, who criticized his Democratic opponent, Lyndon Johnson, for not doing enough to defeat the Communists in Vietnam; Johnson responded by portraying Goldwater as someone who was ready to start a nuclear war over Southeast Asia. Republicans similarly supported Korean operations when Democrat Harry Truman was president, and often bashed him for not doing more—in particular for his refusal to drop nukes on China, as suggested by that sensible General MacArthur. What is interesting is that the aggressive and interventionist streak of the Republican Party in the last half of the twentieth century is in fairly strong contrast with Republican positions during the first half of that century.

Back in the late 1930s, powerful Republicans—with plenty of support from the rank and file—were among the strongest voices arguing for an isolationist stand to keep the United States out of a spreading European conflict. One of the biggest organizations pushing the isolationist message was the American First Committee (AFC), which had international aviation icon Charles Lindbergh as its most prominent and visible spokesman. Lindbergh gave nationally broadcast radio addresses urging American to avoid involvement in the "European War," the basic pitch being that if the Old Countries wanted to go another round that was their business. America, on the other hand, should sit this one out. Lindbergh's words struck a chord. At its height, the AFC attracted nearly a million members. Though it drew some strength from merging with the more left-leaning Keep America Out of the War Committee, the AFC was primarily a product of the American right, particularly the Republican Party, and it was founded by a Yale Law School student, Robert Douglas Stuart, Jr., heir to the Quaker Oats Company, who in the 1980s would be appointed by Ronald Reagan as ambassador to Norway.

This isolationist stance was not surprising—Republicans had taken this as the party line for decades. Twenty years earlier, just after World War I, Republicans played a key role in keeping the United States out of the League of Nations. Powerful Republicans like Henry Cabot Lodge, Chair of the Senate Committee on Foreign Relations, feared the prospect of future European entanglements and wanted no constraints on American sovereignty. Two decades on, many Republicans were singing the same isolationist tune. United States Senator Robert Taft, for example, one of the most powerful and best-known Republican Party figures of the twentieth century, adamantly opposed supporting the allies even as Hitler's armies swept across Europe.[17]

That decades-long isolationism of the Republican Party went out the window as a result of World War II and never really came back. Senator Taft, who in 1939 favored avoiding any involvement at all in a spreading war in Europe was, in 1951, in favor of bombing China in support of a mainland invasion by Chiang Kai-shek's Chinese nationalist forces.[18] In the decades since World

War II, interventionist policies consistently received higher support from Republicans—sometimes, as we have seen, even from the same Republicans who had once been strict isolationists. Fortress America isolationists becoming Cold Warrior interventionists—where is the unifying thread in that?

These patterns only make sense if viewed from the perspective of bedrock dilemmas such as the appropriate relationship with out-groups. The issues of the day—whether to help take on the Nazis, saber-rattle at the Russkies, or pursue Saddam Hussein into his rabbit hole—are shaped by all sorts of framing and partisan effects, contextual factors like who is in power (Democrat FDR or Republican George W. Bush) and the broader social milieu created by, say, the Great Depression or the aftermath of 9/11. Constant throughout these issues, however, is the right's strong belief (relative to the left) that security is paramount and out-groups, whether they be Nazis, Europeans in general, Commies, Cubans, Muslims, or just vague and unnamed evil-doers, should be treated as potential threats. While that orientation resolves the dilemma—people who are not us are potential threats—it does not resolve the *issue* of how to deal with those potential threats in a way that maximizes protection of the in-group. One option is to just wall ourselves off and keep the "others" out. This was just the strategy embraced by Lindbergh and many on the right—but then on December 7, 1941, the Japanese Air Force demonstrated that minding our own business did not always keep the out-groups out. How to deal with these threats, then? The answer, in one form or another, was: Get them before they get us. Conservatives will only support nation-building when it is framed as something that can keep the United States safe by stabilizing a previously dangerous foreign entity. For their part, liberals will find nation-building appealing when it is framed as invoking national self-determination, the welfare of the people of that foreign entity, and integration with the international community.

Here is another example of framing affecting the issues of the day, but doing so against the backdrop of a stable bedrock issue. Exit polls from the 2012 election in the United States indicated that Latinos voted Democratic by better than a 2 to 1 margin. In early 2013, key leaders of the Republican Party, still stinging from their unpopularity with this quickly growing demographic, proposed a path to citizenship for a portion of the estimated 11 million individuals currently in the country illegally. Previously, such a proposal would have been a nonstarter—dismissed as "amnesty," a dreaded word for many Republican primary voters. Electoral incentives can be strong, however, and many Republicans are beginning to view those 11 million (as long as they have no criminal record, have paid all their taxes, have been here a long time, know U.S. history and civics, and speak English) as potentially part of the "in-group." To be sure, suspicion of out-groups remains strong among many conservatives, as is indicated by their accompanying demand that the border first be strengthened in order to keep out any more immigrants. Still, the softening of position indicates that definitions of the in-group and out-groups can be reframed in some circumstances. Does this mean that, compared to liberals, conservatives on average are less suspicious of out-groups? Not in the slightest.

Understanding the unity of politics thus requires diving beneath the issue stances of the day and the vocabulary employed so that it is possible to identify the different sides on those issues. It is at this deeper level that we find a set of predispositions toward social life that are as constant as the force of opposite magnetic poles, pushing people together or pulling them apart regardless of what issues or labels are in play. From this perspective, in order for a universal element of politics to exist, labels do not need to be constant across time and space and neither do issues of the day. Conservatives may have advocated different strategies before and after World War II, but their wariness of out-groups and their aversion to being pushed around by (and potentially made subservient to) out-groups has never wavered. Disputes over bedrock political principles occur always and everywhere since people in all societies differ in their inherent predispositions toward, for example, the nature of leadership or the necessity of adhering to traditional values. What we call these people is beside the point—the point being that differences regarding bedrock dilemmas have existed as long as human beings have been living in social groups.

Small groups—the families, tribes, and bands in which we lived for much of human history—were so intimate and personal that collective decision-making could be sorted out through relatively simple institutions like kin group dominance hierarchies. This sort of social intimacy, though, does not describe politics on the mass-scale. The population of Athens numbered somewhere in the low tens of thousands, and they managed to keep politics reasonably intimate and social only by defining citizenship so narrowly that it effectively cut most people, including Aristotle as it happened, out of collective decision-making.[19] Soon after, however, the continuing development of mass-scale societies created a very different context for dealing with bedrock social dilem-mas. The tangible and personal gave way to the abstract and impersonal.[20] The issues are no longer disputes over the best side of the river on which to camp, or what to do about Zug, who seems to be in the habit of taking more than his share of mammoth meat. Now we are talking about conservatives in Kansas who are trying to get a federal government in Washington, D.C., to prevent two guys in California—two guys they will never meet, two guys who will never meaningfully intersect with, let alone tangibly affect, their lives—from marrying each other. We are talking about liberals who want those same con-servatives in Kansas to cough up tax payments so the same federal government in Washington, D.C. can force people in Utah to buy health insurance poli-cies they do not want. On that scale we are making the leap from the merely social to the truly political. Our preferences regarding bedrock dilemmas—adherence to traditional values and redistribution of resources—are now split-ting us into identifying with big abstract ideas like conservatism or liberalism or whatever ism is popular in your particular culture. Time to pick an alcove and design a tea towel.

Complete This Sentence: Society Works Best When . . .

If our argument for the universality of politics is correct, we should, at least in theory, be able to measure preferences on bedrock dilemmas and these prefer-ences should line up with political attitudes and beliefs in *any* given historical or cultural context. Historically speaking, we can certainly dig up some anec-dotal evidence to support our argument. Aristotle was kind enough to provide some of this sort of thing, pointing out that city states like Sparta and Ath-ens differed crucially in their preferences for leadership styles and collective decision-making, resulting in differing institutions (a monarchy, the assembly), that in turn perpetuated advocacy for those preferences. Such differences did not just show up between ancient polities, but also within them. The late Roman Republic (circa the century before the birth of Christ) was marked by an ideo-logical divide over bedrock dilemmas. The sides were not called conservatives or liberals, but *optimates* and *populares.* The optimates ("best men") wanted to preserve the republic's traditional values and way of doing things, which for practical purposes meant keeping power concentrated in the hands of a wealthy elite and avoiding rule by noisome Julius Caesar dictator types and, especially, rule by even more noisome commoners. The populares ("favoring the people") came from the same aristocratic class as the optimates but were basically populists, supporting welfare programs (subsidized grain for the poor), limitations on slavery, and expanded citizenship rights.[21] Sound familiar? Teddy Kennedy would have looked good in a populare toga. For all intents and purposes, we probably *could* call optimates conservatives and populares liberals.

Ultimately, what divides Athenian and Spartan, Imperialist and Republican, Roundhead and Cavalier, Federalist and anti-Federalist, monarchist and revolutionary, Bolshevik and Menshevik, partisan and Fascist, Alcove No. 1 and Alcove No. 2, Buckley and Vidal, Sarah Palin and Hillary Clinton, Jean-Marie Le Pen and François Hollande, Western democrats and Islamists seeking a new Caliphate are different perspectives on the proper way to design, structure, and maintain society. The underlying tectonic plates may go by different names, but the fault lines between them are uncannily similar.

Along with a number of other researchers, we have been arguing in academic journals for several years that individuals have core preferences on fundamental issues such as leadership, defense, punishment of norm violators, devotion to traditional behavioral standards, and distribution of resources. It is one thing to make a theoretical argument, though, and quite another to provide evidence for it. If the argument is correct, quite independent of labels it should be possible to get a notion of whether someone prefers a society to be run with an assertive leadership style or a society that upholds traditional, unchanging norms of conduct. We should, in other words, be able to tap directly into the universality of politics.

As far back as the 1960s and 1970s, scholars were investigating whether common sets of preferences on things like property rights/resource distribution, egalitarianism/hierarchy, and traditional values/social change and innovation existed across cultures. One political scientist, J. A. Laponce, conducted a large multinational analysis that included representative countries from Europe, North America, Africa, and Asia, and found that politics falls pretty predictably into a more or less universal left-right spectrum.[22] More recently, other researchers have argued that humans have core sets of values or moral foundations that consistently order their differences on politics—something that is not too far removed from what we are arguing.[23]

Even the spatial metaphor—left and right—runs deeper than typical accounts aver. Most humans are right handed and lefties were viewed with suspicion for a very long time. As a result, the left-right opposites metaphor was readily available as an organizing device for social relations. For millennia and in most cultures, the right has been associated with religious and social orthodoxy, the just, and the good, while the left has been associated with the opposite. There is a reason we seek to be righteous and not lefteous. The seating arrangements at the Estates-General were not arbitrary after all, and it is no big mystery why the upper crust was on the king's right. The big exceptions to the left-right social metaphor are the rare societies dominated by left-handed people (where the left is associated with religious and social rectitude) and certain other societies (like the Chinese) who have used more of an up-down duality, with the celestial guardians of social orthodoxy at the top. Even here, though, it is the directional labels that are different and not the bedrock of politics.[24]

Still, while supportive, this research has not explicitly gone looking for bedrock dilemmas in the sense of trying to identify and measure exactly what these critters are. We took a first crack at doing exactly this a few years ago by developing what we called the "Society Works Best" index. The basic idea was to try to come up with a reasonably comprehensive list of bedrock social dilemmas that should make sense regardless of historical or cultural context. Our first attempt was pretty simple; we derived a set of questions that asked people to pick one of two options for fourteen questions that all began, "Society works best when . . . ." The response choices were things like, "Our leaders stick to their beliefs regardless" or "Our leaders change positions whenever situations change;" "People realize the world is a dangerous place" or "People assume all those in far-away places are kindly." Table 2.1 gives a complete list of these questions.

We used these questions to create a simple scale of preferences on social organization. The first time we employed this scale (in 2007 on a sample of 200 U.S. adults) we did not quite believe the results—the Society Works Best index predicted issue attitudes, ideological self-placement, and party identification with astonishing accuracy.[25] The correlations between the Society Works Best index and these other standard measures of issues and labels were consistently around 0.60. Just as a reminder—we are in a discipline where correlations half that size are reasons to click your heels, gloat at conferences, and dream of Nobel nomination committees. Or at least ask for a raise.

Table 2.1 The Original "Society Works Best" Index

Bedrock Social Dilemma 1: Degree of Adherence to Traditional Values/Moral Codes
Society works best when . . .
1-People live according to traditional values
2-People adjust their values to fit changing circumstances

Society works best when . . .
1-Behavioral expectations are based on an external code
2-Behavioral expectations are allowed to evolve over the decades

Society works best when . . .
1-Our leaders stick to their beliefs regardless
2-Our leaders change positions whenever situations change

Bedrock Social Dilemma 2: Treatment of Outgroups/Rulebreakers
Society works best when . . .
1-People realize the world is dangerous
2-People assume all those in far away places are kindly

Society works best when . . .
1-We take care of our own people first
2-We realize that people everywhere deserve our help

Society works best when . . .
1-Those who break the rules are punished
2-Those who break the rules are forgiven

Society works best when . . .
1-Every member contributes
2-More fortunate members sacrifice to help others

Bedrock Social Dilemma 3: The Role of Group/Individual
Society works best when . . .
1-People are rewarded according to merit
2-People are rewarded according to need

Society works best when . . .
1-People take primary responsibility for their welfare
2-People join together to help others

Society works best when . . .
1-People are proud they belong to the best society there is
2-People realize that no society is better than any other

Society works best when . . .
1-People recognize the unavoidable flaws of human nature
2-People recognize that humans can be changed in positive ways

Bedrock Social Dilemma 4: Authority and Leadership
Society works best when . . .
1-Our leaders are obeyed
2-Our leaders are questioned

Society works best when . . .
1-Our leaders call the shots
2-Our leaders are forced to listen to others

Society works best when . . .
1-Our leaders compromise with their opponents in order to get things done
2-Our leaders adhere to their principles no matter what

Since then we have modified the questions a bit and posed them to a number of other groups of people.[26] These groups include two different twin samples— one in Australia and one in the United States—thereby allowing us to establish that the preferences measured by the Society Works Best index are heritable and capable of predicting issue attitudes across different countries (not all with correlations of 0.60, mind you, but consistently positive and significant).[27] Most recently we included a revised Society Works Best index in a study of 340 adults randomly selected from a particular county in the Midwestern United States. Figure 2.1 plots scores on this modified index, with higher values indicating what could be thought of as conservative positions (traditional values, unbending leaders, punishment of rule breakers preferred to rehabilitation, and rewards assigned on the basis of merit rather than need) against preferences on issues of the day (the previously described Wilson Patterson index), again with conservative positions assigned higher numbers. Figure 2.1 passes what some statistical types call the "inter-ocular shock test." In other words, no fancy

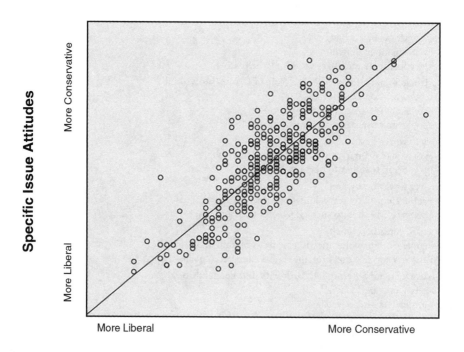

Preferences for Social Order

Figure 2.1   Specific Issue Attitudes and Preferences for Social Order

statistics are needed to see that there is a relationship because it just jumps out and hits you between the eyes. Conservative positions on bedrock principles predict conservative positions on issues of the day, liberal bedrock principles predict liberal issue stances, and moderate positions on bedrock principles predict moderate issue stances. Though much more work needs to be done in a broader array of country contexts, the various versions of this index appear to be fairly predictive not only of issue attitudes but of partisanship and self-reported ideology.[28]

If we accept this idea of bedrock dilemmas as having at least some credence, we also have a basis for taking seriously that biology might be relevant to mass-scale politics. Many people—with some justification—find the notion that there might be genes, particular quirks of neural architecture, or physiological functions that increase the probability of an individual favoring the Bush-era tax cuts or opposing a shift from direct farm subsidies to cut-rate crop insurance to be a bit far-fetched—and so do we. The notion that specific issue attitudes might be biologically instantiated can seem downright silly since issues clearly are tied to culture and the vagaries of historical circumstance. We hope to have explained how ephemeral issues of the day rest on the foundations of universal bedrock social dilemmas. Indeed, we think the behaviors and attitudes driven by preferences with regard to bedrock social dilemmas are likely as primitive and nearly as powerful as other motivators of social attitudes and behaviors widely recognized as being influenced by our biology—things like the need for reproduction and sustenance.

Conclusion: Ideologies 'R' Us

Humans have always brought order to their world by thinking in terms of opposites—light and dark, hot and cold, good and bad, tall and short. All known human societies have used these sorts of classifications and even the most primitive cultures have structured their social relations by thinking in terms of opposites.[29] Sometimes this duality is contrived, but in the case of politics it is not. The division is real and it is unavoidable, and it centers on distinct orientations to mass-scale social life that are typically called ideologies. Ideology is not, as the "end of ideology" school asserted, a concept that just popped out fresh and new from Renaissance thought, only to fade from sight with the end of the Cold War.[30] Ideology is not, as Converse and his many followers claimed, merely the ability to describe currently popular labels or to endorse clusters of positions that meet with the approval of political scientists. Ideology is us. It could no more "end" than could personality. It could no more be restricted to societal elites than could interpersonal communication. Context-specific issues and labels often consume attention and energies to the point that we are blinded to the underlying bedrock principles involved. Debates about capital punishment are context specific; debates about the appropriate treatment of in-group members who have violated social norms are as enduring as bedrock.

Having established that the nature of politics, despite the protestations of many, is universal, our task now shifts to the nature of the human condition. We assert that there is no singular human nature but that individual humans have distinct natures (or predispositions). Important variations from person to person exist and are responsible for the dangerously volatile nature of politics. These variations are the subject of the next chapter.

Notes

[1]   Gross, "The Indoctrination Myth."

[2]   Dewy and Kallen, eds., *The Bertrand Russell Case*. Russell won the Nobel Prize for Literature in 1950.

[3]   Wald, *The New York Intellectuals: The Rise and Decline of the Anti-Stalinist Left from the 1930s to the 1980s*. See especially the portrait of Irving Howe, 311–320.

[4]   Kristol, "Memoirs of a Trotskyist." For those not up on their Marxist history, a Trotskyist is a follower of the theory of Marxism advocated by Leon Trotsky; Trotskyists' main rivals were Stalinists and they disagreed vehemently about things like proper way to bring about a dictatorship of the proletariat.

[5]   Stelzer, "Irving Kristol's Gone—We'll Miss His Clear Vision."

[6]   Laertius, "The Lives and Options of Eminent Philosophers: Life of Aristotle." Trans. C. D. Yonge.

[7]   As the discussion in the text indicates, there is some debate about the exact translation. Aristotle seems to have used the phrase to express at least three distinct ideas. See Mulgan, "Aristotle's Doctrine That Man Is a Political Animal."

[8]   This translation comes from Barker, *The Politics of Aristotle.*

[9]   Mulgan, "Aristotle's Doctrine That Man Is a Political Animal."

[10]   Benjamen et al., "Genetic Architecture of Economic and Political Preferences."

[11]   Alford et al., "The Politics of Mate Choice"; and Stoker and Jennings, "Political Similarity and Influence between Husbands and Wives." On online dating see Huber and Malhotra, "Political Sorting in Social Relationships: Evidence from an Online Dating Community."

[12]   Alford et al., "The Politics of Mate Choice."

[13]   Xiaohe et al., "Social and Political Assortative Mating in Urban China."

14   Heywood, *Political Ideologies*.
15   Zaller, *The Nature and Origins of Mass Opinion*.
16   Shils, "Authoritarianism: Right and Left"; *Lasch, The True and Only Heaven;* and Converse, "The Nature of Belief Systems in Mass Publics."
17   Taft was once described as having "the psychology of the moat." For a contemporary, not necessarily flattering, portrait of Taft's foreign policy, see Armstrong, "The Enigma of Senator Taft and American Foreign Policy."
18   Though to give Taft his due, he never managed to stray far from his isolationist roots. In his last speech, given just a few months before he died, he argued against sending American troops to China or Indo-China. There have also been high-profile Republicans who have continued to champion the isolationist line—Pat Buchanan, for example. But remember, we are talking about tendencies and probabilities; not every Republican supports every military intervention.
19   Aristotle was not a citizen but rather a metoikos or "metic"; this basically meant he was a resident alien.
20   See Diana Mutz, *Impersonal Politics*.
21   Coleman, *A History of Political Thought,* 241–243.
22   Laponce, *Left and Right: The Topography of Political Perceptions*.
23   See Chapter 3 for an in-depth discussion of this research.
24   For an excellent discussion of the history and development of the left-right metaphor in politics, see Laponce, *Left and Right: The Topography of Political Perceptions*.
25   Smith et al., "Linking Genetics and Political Attitudes: Reconceptualizing Political Ideology."
26   For example, we expanded the response categories to standard five-point Likert scaling and altered the question stem from "Society works best when . . ." to "Thinking about politics . . . ." Further refinements are in the works.
27   The results for the Australian sample were weaker, but we think that is due to the fact that in that sample we used a very early version of the social principles index that had not been pretested. Even so, the social principles index was still statistically and significantly correlated with a broad index of issue attitudes. That attitude index included some very different issues from those we asked about in U.S. samples (aboriginal land rights, for example).
28   Funk et al., "Genetic and Environmental Transmission of Political Orientations."
29   Cornford, *From Religion to Philosophy*.
30   See Bell, *End of Ideology* and Fukuyama, *The End of History and the Last Man,* but also Jost, "The End of the End of Ideology."

CHAPTER 3

There is No Normal

Every normal person, in fact, is only normal on the average. His ego approximates to that of the psychotic in some part or other and to a greater or lesser extent.

Sigmund Freud

It may be normal, darling; but I'd rather be natural.

Truman Capote

Stanley Milgram is famous for teaching us that getting people to do really, really bad things to other people does not require an evil demagogue with a Charlie Chaplin mustache. All you need is five dollars and a lab coat. No Gestapo loomed over the good citizens of New Haven, Connecticut, who were recruited by Milgram's lab in the late 1950s. These folks were given a token participation fee and politely asked by a mild-mannered social scientist to inflict potentially devastating harm on innocents. They were alarmingly quick to comply.

It was all a sham, thank goodness. Milgram was interested in who would obey authority figures and under what conditions—pretty big questions for social scientists at a time when one world war had been triggered by authoritarian Fascists and another threatened by totalitarian Communists. Maybe there was something about human nature that led people to obey. Would Americans chuck their morals and act atrociously toward other people just because some authority figure told them to? To answer this question, Milgram set up an experiment that led his participants to believe they were being asked to send increasingly strong electric shocks into another experimental subject as punishment for failures in a word-pairing exercise. In truth, the individual ostensibly trying to remember the word pairs was an experimental confederate who was not being shocked at all.[1]

Milgram's experiments are among the most famous in all of social science and their most famously disturbing finding is that even when people thought they were being asked to shoot more than 400 volts of electricity into a fellow participant who had done them no harm, they often complied. Milgram's results led to much handwringing about the human condition. If homo sapiens are this willing to obey, it is no surprise that otherwise decent human beings follow the edicts of evil authority figures. The depressing, even frightening, implication was that given the right circumstances, we would all become complicit in the perverted policies of a Himmler or an Eichmann.

Other experiments seemed to back up this conclusion. The Stanford prison experiment, conducted in the 1970s by psychologist Philip Zimbardo, is nearly as famous as Milgram's in suggesting that we all have a hidden Nazi somewhere in our souls. This experiment involved randomly assigning college students to play either inmates or guards in a mock prison set up in the basement of Stanford University's psychology building. The "prisoners" were "arrested" (with the cooperation of the local police) and hauled off to the prison, where the "guards" were decked out in uniforms and mirrored sunglasses. Without being told to, the guards quickly began behaving badly, humiliating their charges and engaging in various forms of psychological abuses, such as stripping prisoners naked, putting them in solitary confinement, and forcing them to sleep on concrete floors. Remember: The only crime of the "prisoners" was to be assigned arbitrarily to play the role of prisoner. Nevertheless, the guards quickly began to abuse their authority; indeed, some displayed what can only be described as sadistic tendencies. The broadly accepted take home point of research like Milgram's and Zimbardo's is that *everyone* is capable of behaving inhumanely if the conditions are right.[2]

Except they are not.

Let's go back to Milgram's lab, where good American burghers were lighting each other up just because a guy in a lab coat asked them to. Milgram's ersatz shock machine was calibrated in 15-volt increments, beginning with 15 and going to 450, and his primary interest was the number of participants who would go the full monty. He found that the likelihood of participants cooperating all the way to the 450-volt switch was increased by a number of environmental factors. Notably, he varied whether the subject controlling the shock box could hear or see what was being done to the shockee. When the poor schlub getting jolted was neither seen nor heard by the individual in control of the levers of power, 65 percent of the participants were willing to go to 450 volts; when the recipients' discomfort was heard but not seen, obedience dropped to 62.5 percent; when it was both heard and seen, full obedience plummeted to 40 percent. When Milgram's research participants were required to force the protesting subject's hand onto the contact plate to get jolted, only 30 percent of the participants went all the way. Averaging across Milgram's four core experimental manipulations, a bit more than half—81 of 160—of the participants refused to go all the way to 450 volts.

It is hardly surprising that people are less willing to inflict pain on someone they can hear, see, and feel, so the variation across these experimental manipulations was not seen as a serious caveat to the central conclusion that humans are inclined to obey authority figures. Typically passed over—certainly by Milgram himself—is a far more intriguing finding; namely, the remarkable variation in the behavior of his participants within the *same* situation. Even when the recipient was only an abstraction—neither seen nor heard—more than a third of the experimental participants refused to go to 450 volts. Some refused as early as 105 volts, when the experiment was just getting started. When the official, scientifically attired Yale scientist said, "The experiment requires you to go on," one in three people essentially said, "Forget it, I'm not going to hurt that guy." Conversely, even when the research participant had to place the learner's hand on a contact plate to complete the circuit, nearly one out of three was still willing to go all the way. Makes you wonder under exactly what circumstances these particular individuals would *not* obey authority figures.

Our point is this: Milgram's research is typically invoked as evidence that all humans are capable of atrocities, but its real message is that some of us are and some of us are not. If his sample is representative of the rest of the population, Milgram's results suggest that roughly one third of all people are strongly inclined to obey authority no matter the painful implications for innocent others;

another third are obedient in some circumstances and resistant in others; and the final third are commendably resistant to authority figures when the fate of innocents is at stake.

Milgram's results indicate the existence of a great deal of individual-level variation in obedience. Zimbardo's prison experiment similarly found wide variance in the tendency of participants to turn into "evil guards." Some did and some didn't. These findings square with outside-the-lab reality. Not everyone fell in line behind Hitler and Mussolini; not all young Americans defied authority in the 1960s; and not all Chinese were at Tiananmen Square on June 5, 1989, let alone standing calmly in front of a row of tanks. Each of the three of us has raised multiple offspring and that experience tells us that children—even children raised in very similar environments—vary wildly in their tendency to obey authority figures such as parents.

At one level, this marked variation in the degree to which individuals defer to authority, like the marked variation in virtually all behavioral traits, may not be surprising—yet social science research often undersells, or even ignores, individual variation in behavioral predispositions in favor of variations from situation to situation. This chapter is all about individual differences and why they are important.

Some for You and Some for Me

Milgram is far from the only social scientist to play down the remarkable person-to-person variation in human behavior. In fact, assuming that people are all the same down deep is pretty much the norm. As an example, consider research on people's tendency to share with others and to punish those who do not share. Social scientists often study these sorts of traits with variants of "divide the dollar" games. The most basic of these games works like this: A social scientist randomly grabs you and someone you don't know, then hands you 20 crisp one-dollar bills and declares you are free to share none, some, or all of your newfound largesse with the stranger. You make your decision, the money changes hands—or not—and everyone goes on their way with their allotted sum. This is called a dictator game because you have dictatorial powers and the stranger has to take what you give, even if what you give is nothing. A well-known variant is the ultimatum game. Here the stranger must approve of how you propose to divvy up the 20 bucks. If the stranger vetoes your proposal the social scientist takes back the money and you both get zippo.

Classical economic theory has very precise predictions about what you will do with the money. In the dictator game you will not give the stranger anything. Why should you, since you will probably never see him or her again? The rational thing to do is to maximize your benefit and that means holding onto the fistful of dollar bills. You cannot be similarly Scrooge-like in the ultimatum game, though, because the stranger has a veto. Give the stranger nothing and you are likely to get nothing. The problem is how much to give. Economic theory predicts that you will give the least amount required to avoid a veto. If you are holding 20 one-dollar bills, that amounts to a measly dollar. Here's the logic: Walking away with a dollar is better than walking away with nothing, so a dollar should be enough to prevent a rational stranger from exercising a veto.

These sorts of games have been repeated thousands of times in an amazing variety of contexts, and with an amazing variety of twists and minor modifications. The clear message from all this research—a message that is surprising only to economists—is that classical economic theory stinks at predicting how people will divide their 20 dollars. People are wildly more generous to strangers than they need to be. The average amount passed along in a dictator game is not zero but rather about $8 of the $20; in other words, pretty close to an even split and way more than rational maximizing behavior would suggest.

The results of ultimatum games are even more interesting. Remember, a rational person should accept any positive amount because one dollar is more than no dollars. In reality it is very common for small offers to be rejected. If you keep $19 and offer just $1, many strangers will exercise their veto and your 19 bucks will go poof. Splits of $18–$2, $17–$3, $16–$4 also are frequently turned down; even $15–$5 splits are occasionally nixed. What all this tells us is that people routinely deviate from rationality in order to be generous to a powerless stranger or to stick it to a greedy bastard. These findings probably are not big news to you but they create serious problems for the theory that humans are rational maximizing actors because, well, they don't seem to act very rationally.

This basic message stays the same even when researchers tinker with the setting or format of the basic script. These games have been played in Siberia, in Western universities, and in hunter-gatherer societies.[3] The stakes of the

games have been altered by taking them to regions of the world where $20
is the equivalent of several months' wages.[4] The $20 has been described as
a blind (an unseen resource) or a pot rather than as a fund belonging to the
divider.[5] The physical attractiveness of the "stranger" has been altered.[6] And
the "stranger" has been rendered less strange by altering the extent to which
the players know each other.[7] These changes make a difference, driving non-
maximizing behavior up or down, but none alters the basic conclusion that
people are not the single-minded pursuers of profit that economic theory
holds them out to be.

Just as Milgram's results are presented as indicating that people are subservient to authority, the divide the dollar outcomes are presented as evidence
that people are irrational; and just as the common interpretation of Milgram's
research is mistaken, so too is the common interpretation of the research
on economic games. A closer look at the game results indicates tremendous
individual variation in the decisions people make—even when the locale and
experimental manipulations are the same. Some people are simply more generous than other people; some are more punitive; some are more strategic; some
are more consistent; and some are more sensitive to the setting.

A significant minority of people—our best guess is around 20 percent—play
economic games in a manner that is quite consistent with classic microeconomic theory in that they do not share unless they have to and they do not
punish those who do not share with them. Others are relentlessly generous
and the decisions of still others are variable and contingent upon context. The
common conclusion growing out of the economic games research—that people are not rational maximizers—badly misses the point. Whether the topic
is obeying authority figures or sharing resources with strangers, the real message of empirical research on human behavior is that people are fundamentally
different. "People" are not lemmings in the face of authority—but some are.
"People" are not rational maximizers—but some are.

Lake Situationalist: Where All the Children Are the Same

So why do individuals in exactly the same situation behave so differently?
Milgram expressed little curiosity about this question, saying he "left to other
investigators" the task of studying variation across individuals. In his major

book on the obedience experiments, he devoted a sum total of three pages to individual variation. His central conclusion was that those obedient to authority saw the learners as responsible for their predicament (the learners received a shock because they failed to match the word pair) whereas subjects defiant of authority saw themselves as responsible (the learners received a shock because I flipped a switch). He ducks the obvious and more interesting question of why participants view responsibility so differently. Instead, he makes a half-hearted attempt to look at demographic variations across the defiant and the obedient. Finding few differences (except for level of education), Milgram punts and concludes with what could be the motto of much of social psychology: "[I]t is not so much the kind of person a man is as the kind of situation in which he finds himself that determines how he will act."[8] In short, Milgram had no clue why some people in exactly the same situation eagerly turned on the juice while others refused. He certainly never entertained the possibility that obedient and disobedient participants might be physiologically, cognitively, psychologically, and genetically different.

Milgram's focus on the situation as the key explanation of behavior and his abject indifference to behavioral variation within the same situation is disconcertingly typical of social science research. As an illustration of the value that could be added if this research tendency were altered, consider a fascinating study conducted some time ago by economist Kevin McCabe and colleagues. They had participants play a variant of divide the dollar games called a "trust" game while their brains were being imaged. The twist in this case was that players sometimes interacted with another human being and sometimes with a computer that was programmed to follow a preset sequence. McCabe found that people's brain activation patterns are quite different in these two situations.

Told they are playing a computer, little activity registered in the emotional (or limbic) areas of the brain or in the prefrontal cortex of participants. In this situation the brain appears to be on autopilot, doing nothing more than calculating the way to get the most money (in other words, to be rational). Against a human being, in contrast, limbic areas such as the amygdala are activated, as is the prefrontal cortex, which presumably must resolve the conflict created by the rational desire to acquire more money and the emotional feelings that might accompany an exchange situation.[9]

If it ended there, this research would be another example of the kind of approach that we are cautioning against: general statements that "people" display different brain activation patterns depending on the situation. This particular study, however, has a feature that illustrates the value of looking at individual differences. When the five most uncooperative individuals, as determined by the decisions they made in earlier economic games, were observed in the scanner, their brain activation patterns, unlike other participants, tended to be no different when they were playing against another human being than when they were playing against a computer. Thus, at least some people appear to be surprisingly devoid of the emotional responses that typically accompany human interaction.[10]

Tellingly, this powerful and provocative finding was relegated to a couple of brief references in the published study, but the authors should be given full credit for paying even a little attention to this sort of result. The usual study would have aggregated the neural patterns of all participants on a situation by situation basis, made global statements about people's brain activation patterns in those situations, and left it at that. By considering separately those who are and are not predisposed to cooperate, doors are opened to all kinds of possibilities. The important question is not just what situations make people more or less cooperative; it is why some but not other people are limbically muted noncooperators.

The bread and butter research design of the social sciences is to identify a situation of interest, find (through the historical record) or create (through experimental manipulation) scenarios where that situation is and is not present, and measure the behavioral differences between people in that situation and people who are not in that situation. Don't get us wrong. This approach is valuable—but the obvious next step is to identify the people for whom the pertinent conclusions are more or less applicable. To take an example from research on optical illusions, we know that, on average, people are slightly more likely to see two dots as widely spaced when the dots appear inside an object than when they appear on their own. We know much less about those individuals who defy this tendency to engage in what is known as "object warping."[11] What kinds of people are more likely to have their judgment of distance affected by nearby objects? And who knows, some people may even go in the opposite direction, claiming greater distance for dots in open space than those in a rectangle. In comparing the average behavior *between* situations, the remarkable variation in behavior from one person to another in the same situation is all too often ignored.

There's More Than One Way to Skinner a Pigeon

Theory is one of the big reasons that behavioral variation within a given environmental condition is often the stepchild of the social sciences, and we will show how each of the major social science theories ignores important individual variation. Let's start with "behaviorism," a conceptual framework that dominated social science research and thinking for a big chunk of the twentieth century. Probably the best known behaviorist was psychologist B. F. Skinner, who believed most organisms could be conditioned to do just about anything, given the right rewards and punishments. Pigeons could be made to press levers, rats to negotiate mazes, and humans to hate the color purple.

In strict behaviorism, the interior architecture of an organism is irrelevant; all that matters are the carrots and sticks that direct the organism's behavior in one direction or the other.[12] Meaningful differences across pigeons in baseline lever-pressing aptitude were simply dismissed by behaviorists; a pigeon deficient in lever pressing is by definition a pigeon insufficiently exposed to the appropriate rewards and punishments. According to behaviorists, it doesn't matter if the species of interest is pigeons, rats, chimps, or humans—organisms subjected to exactly the same rewards and punishments throughout the entirety of their lives will behave in the same fashion. The environment—that is, the situation—is wholly deterministic. Skinner's teacher, John Watson, famously boasted, "[G]ive me a dozen healthy infants . . . and my own specified world to bring them up in and I'll guarantee to take any one at random and train him to become any type of specialist I might select—doctor, lawyer, artist, merchant-chief, and, yes, even beggar-man and thief."[13] The notion that deep-seated predispositions might meaningfully affect the ability of situations to mold behavior simply was not taken seriously by behaviorists.

Behaviorism fell out of favor when it became clear that innate inclinations did indeed mold behavior. Psychologist Harry Harlow's work with rhesus monkeys famously demonstrated the importance of such predispositions. Harlow showed that baby monkeys preferred to spend time with a cuddly surrogate mother even when a decidedly noncuddly surrogate mother provided a tangible reward that the cuddly mother did not: milk. No amount of training and environmental manipulation could divest the baby monkeys of their desire to be in contact with the soft, non-milk-giving doll. The failed attempt at conditioning, though, did have a big impact on behavior: Baby monkeys deprived of contact with the cuddly surrogate grew up to be socially dysfunctional adults.[14]

A couple of years before Harlow's monkey experiments, a scientist named John Garcia was studying the effects of ionizing radiation on rats. Garcia noticed that lab rats would not drink from a plastic bottle, presumably because of the taste. This led Garcia to start experimenting with taste aversion. He found that if rats drank foul-smelling (but actually clean) water and then were made sick (by radiation) they would not drink foul-smelling water again. Garcia repeated the experiment where the water smelled fine but was bubbly and unnaturally colored. No matter how many times Garcia made the rats ill after drinking the pastel bubbly, they could not be trained to avoid it.

That result did not fit with behaviorism at all. Why were rats so easily conditioned by water that smelled bad but not by water with a bizarre, artificial appearance? The obvious answer is that evolution had selected rats that avoided rancid-smelling water because those that did not often died. No cave rat, though, ever encountered effervescent, pastel-colored water. Lacking this evolutionary history, modern-day rats are left wholly incapable of connecting the peculiar-looking water with their subsequent illness.[15] Garcia's finding is a clear violation of the cherished behaviorist principle of "equipotentiality," the ability of all stimuli to serve as equally powerful conditioning agents. Instead, organisms appear to possess preexisting behaviorally relevant dispositions toward certain situations and stimuli.

Once these first chinks in behaviorism's armor appeared, the collapse was on. Apes raised in captivity that had never seen a snake could nonetheless easily be conditioned to fear them, even though they could not be conditioned to fear other less venomous animals.[16] A range of innate human phobias quickly became the target of several active research agendas.[17] And Darwin's classic observations about the amazing behavioral differences from one breed of dog to another were revived.[18] Behaviorism's assertion that all behavior is learned behavior was battered hard by an ever-growing list of studies demonstrating that some behaviors are not attributable to learning.

Strict behaviorism is now out of fashion in most academic circles, but other widely discussed theories also fixate on the role of the environment in determining behavior. For example, evolutionary psychologists generally take a dim view of behaviorism and are quick to point out that biology matters. Even Skinner, they slyly observe, could not condition rats to fly, pigeons to swim underwater, and chimps to multiply 6 times 7. The necessary biological infrastructure simply isn't there to support these behaviors. Yet despite their recognition of evolutionarily shaped biological and psychological infrastructure, evolutionary psychologists can be surprisingly behaviorist in their belief that it is the environment alone that drives behavioral variation within a species.

A basic assumption of evolutionary psychology is that every species has a "universal, species-typical architecture," which reflects the functional adaptations of that species to natural selection pressures.[19] Consider a standard, Mark I Homo Sapien. It does not matter what historical era or culture your Mark I model comes from, it will reflect the same blueprint. Barring manufacturing errors, damage during shipping, or unusual wear and tear, it will have two legs, two arms, two opposable thumbs, a set of facial expressions for primary emotions (smiles for happy, frowns for sad), and a set of psychological "modules" sensitive to things like detecting other humans who are cheaters.

Universal, though, does not mean inflexible, and this standard architecture is capable of producing extraordinary variation if it gets the right signal from the environment. The most spectacular examples of this sort of flexibility come from nonhuman species. Consider the reef fish known as cleaner wrasse, which typically form social groups of one male and several females. Take the guy out of this equation and something amazing happens—one of the females, usually the largest, will become anatomically male. The wrasse's universal architecture has given it the capability of changing gender if the situational pressures are right. And wrasse are far from the only species that can do this sort of thing. When alligator eggs are initially laid, their sex is indeterminate. Whether the hatchling is male or female is based largely on the temperature of the nest during incubation. Cooler temperatures produce clutches of females; warmer temperatures, males.

While not turning Janes into Joes, the universal architecture of primates, including humans, also permits measurable biological changes based on environmental situations. For example, remove the alpha male from a troupe of chimps and the testosterone levels of the "second-in-command" shoot up; he's literally getting hormonally primed to be top dog (or chimp).[20] The point here is that these organisms are not "hardwired" to be male or to be juiced with a particular level of testosterone. Instead, their underlying genetic architecture permits substantial flexibility for the organism to be shaped by its environment.

Though the notion of universal architecture seems to allow evolutionary psychology to account for behavioral variation, the source of the variation is the situation and not deep-seated, perhaps genetic, biological, individual-level variation. Universal architecture allows flexible responses, but the architecture itself is still universal. Evolutionary psychologists typically acknowledge that architectural differences exist across reproductively meaningful groups such as gender and perhaps age, but otherwise the notion that variation in biological architecture is responsible for variation in behavior is treated skeptically. This is where we part company with evolutionary psychology and all theories (like behaviorism) based on the notion that there is a single human nature.

For example, evolutionary psychologists have been particularly critical of the notion that human personality differences are genetically influenced, adaptive, and anything other than facultative.[21] Unlike behaviorists, the prime situational mover for evolutionary psychologists is not the social scientist, bristling with levers, sugar water, mazes, food pellets, and ersatz shock machines, but rather the environmental situation: the temperature of the nest, the prevalence of pathogens, the emotional climate of the childhood home, the orientation of fellow workers, the availability of necessary nutrients, or the trustworthiness of strangers.

Classical economic theory is in much the same boat. We have already noted this theory's spectacularly inaccurate predictions with regard to various divide the dollar games. Classical microeconomic theory ends up in the same situational place as behaviorism and gets there much faster than evolutionary psychology. This is because it, too, is built on a worldview of presumed human universality, specifically humans as preference-maximizing machines. We might prefer beer and you might prefer wine, but the reasons we have different preferences is not of interest to most economists. They are more excited by the presumed universal process people employ to maximize those preferences in a given situation (rational utility maximization, as it's called in the trade).

Classical economists rarely recognize the relevance of behavioral morphs. While psychologists study introverts and extroverts and political scientists study liberals and conservatives, economists have no parallel widely accepted terms that are indicative of fundamental economic types.[22] The situation determines what people need to do to maximize preferences so there is no need to worry about the fiddle-faddle of people having different preferences in the same situation. Preferences are taken as given (in other words, assumed away), and when deciding what to do, it is assumed that all humans crank through a universal cost-benefit calculation. The perceived pros and cons in that calculation are determined not by variation in personality, or neural architecture, or cognitive processing styles, but by the situation. As Dennis Mueller wisely notes, "homo economicus . . . bears a close resemblance to Skinner's rat."[23] The point is that broad swathes of the most prominent social science theories are based on the assumption that the human condition is monolithic and that any variations in human behavior are exclusively the product of the situation. The problem with this assertion is that it is simply not true.

Locke, Stock, and Gladwell

This tendency to view people as interchangeable and situations as determinative is by no means restricted to hoary social science theories. Authors, philosophers, and public intellectuals also typically explain behavioral variations via context and not dispositional differences. Thomas Hobbes thought human nature was so nasty that we needed an oppressive government to save us from ourselves.[24] John Locke is frequently held up as the light to Hobbes' darkness, but Locke was not much cheerier about basic human nature. He thought people would be nice, but only if conditions made it unprofitable for them to be mean. He pinpointed the technological advances of salting meat and coining currency as creating conditions ripe for meanness. As long as anything valuable was perishable, it made no sense to stockpile goods beyond what could be consumed in the next day or two—so go ahead and have an extra slice of my mammoth meat. Invent money and preservatives and the gloves come off; now if you want a taste of my cured ham, it'll cost you. Locke did not see differences between people as particularly consequential and believed that crucial situational changes in the long-distant past allowed another side of human nature to be manifested.[25]

Karl Marx also believed human behavior was driven by situations—more specifically, by position in the class structure. Owning the means of production led to one sort of behavior, mostly associated with exploiting the proles. Being stuck in the lumpenproletariat led to another sort of behavior: mostly trying to put food on the table. Marx believed that when capitalism was defeated, resources would be properly distributed, scarcity would end, and people would no longer act selfishly. This end stage of the Marxist dialectic can be thought of as mankind emerging from the long, miserable epoch entered into when Locke's precious metals, currency, and barterable goods became part of the picture. In short, Marx believed capitalism alienates both capitalists and the proletariat from their true nature and that if the situations endemic to capitalism could be eliminated, people would also change. So while Groucho Marx believed Republicans and Democrats are fundamentally different, Karl Marx believed a Democrat is just a Republican who owns no means of production. The Marx with the better intuition is clearly Groucho.

Karl Marx's economic determinism parallels the cultural determinism that seduced the likes of Gauguin, Rousseau, Wittgenstein, Isaiah Berlin, Emile Durkheim, T. S. Eliot, and Franz Boas. The "noble savage" movement asserted, contra Hobbes, that humans in the state of nature are good, dignified, and virtuous. The behavior associated with this fundamentally good human nature, the story goes, went off the rails with the advent of cities, congestion, car pools, and nine-to-five clock punching. If we could only change the situation—return people to bucolic settings—they would behave admirably and all would be well.

Durkheim did as much as anyone to shunt scholarly attention away from the individual. Often considered the founder of modern sociology, he was interested in explaining how societies could maintain their integrity and coherence in the wake of modernity. To study this topic he believed that social science should be holistic: It should investigate phenomena attributed to society at large rather than the specific actions of individuals. He lamented "forced division of labor" and was proud of being a structural determinist, someone who believes that behavior is determined by the structure of the social, linguistic, and cultural system in which people are embedded.[26]

One of the more interesting behaviors this group attributed to social structure was sex. Pioneering anthropologist Franz Boas believed that the sexual problems so prevalent in the West were merely the product of effete urbanity. Seeking evidence consistent with this romantic primitivism, he convinced his student, Margaret Mead, to go to Samoa to document the absence of sexual hang-ups in primitive societies. Mead did not speak the language, spent less than nine months on the island, and lived with a western dentist during the entirety of her stay, but famously reported Samoa as a sexual paradise devoid of jealousy, acrimonious breakups, and guilt-ridden infidelity, not to mention rape and suicide. Perhaps not surprisingly, upon further review, Mead's description turned out to be open to challenge, and some evidence suggests Samoans are not quite the free love, no consequences bunch Boas and Mead so fervently wanted them to be. In fact, evidence has been presented that a few mischievous young Samoan women thought it would be fun to put one over on Mead, and so painted an inaccurately idyllic picture even though the women themselves were intimately familiar with the unseemly side of Samoan sex life.[27]

One of the most tragic consequences of the noble savage mindset occurred in China. Chairman Mao thought the uncorrupted rural peasant embodied all that was good about the human condition, so he tried to create more of them. He forced millions to relocate from China's teeming cities to its pastoral environs. This "make-a-peasant" program was, to say the least, a failure. Instead of thriving, the transplants died in droves—upwards of 20 million by some estimates. The view that social context alone determines human behavior—that individual variation does not matter—has been a source of misunderstanding and even catastrophe throughout history.

The tradition of dismissing meaningful individual-level human variation is not restricted to philosophers, Communists, and devotees of the noble savage concept. It can also be found on modern best-seller lists. Take, for instance, the work of Malcolm Gladwell.[28] In one book, Gladwell says he wants to go "beyond the individual" in explaining why some people are successful and some are not. He writes of "hidden advantages," "extraordinary opportunities," "cultural legacies," and "hard work," and says the keys to success are luck and diligence. Want to be the next dominant hockey player on the planet? Then your best strategy is to have a birthday just after the age cutoff used to classify youth teams. That way you are more likely to be the oldest and most physically mature specimen on the squad, thus increasing your chances of developing confidence, being selected for the traveling squad, getting to refine your skills even more, and going on to be the next Wayne Gretzky. Hockey not your thing? Never fear, the basic principle applies to doing sums, playing soccer, and strumming a guitar.

Gladwell does offer hope for those whose birthdays do not fall at the right time of the year but it has little to do with natural aptitudes and core individual differences. If you want to succeed, all you need to do is practice. Not just practice a little but practice a lot—a minimum of 10,000 hours. The Beatles, Gladwell claims, made it big not because of any particular musical talent but because when they were fledgling musicians they packed themselves off to perform in Hamburg dives where they refined their skills by playing extended shows in front of tough crowds night after night. But so did Tony Sheridan and Rory Storm and the Hurricanes, and not many people have heard of them, though they have probably heard of the Hurricanes' drummer. His name is Ringo and the Beatles poached him after they gave Pete Best the boot. Perhaps Pete slacked off and only practiced 9,999 hours.

Importantly, Gladwell treats the capacity to dedicate yourself to a punishing practice regimen as something that is purely a matter of individual will. The assumption seems to be that any one of us could be the next Paul, John, George, or Ringo because we all possess the willpower to put in 9 to 10 hours every single day for three years on our Stratocasters. That's a pretty big assumption, since the required dedication to practicing a craft simply is not something everyone has. People who do not put in 10,000 hours mastering a single skill, we want to emphasize, are not slackers. Spending all that time in the gym, at the library, or practicing chord progressions to a Merseyside beat means you have to sacrifice a lot. Not everyone has an inner drive so strong they are willing to live potentially unbalanced lives to nurture it. Gladwell's message seems to be, "You too can be great if you just work at it." Our message is that most people are not predisposed to work at it to the degree required to become great. They are not necessarily lazy but physiologically, cognitively, psychologically, and perhaps genetically different from those who are willing to dedicate themselves in this fashion.

The same is true for differences in approaches to lots of other things. As we will go to some lengths to demonstrate in the next chapter, attitudes toward everything from a curiosity about exotic foods to a tolerance of alternate lifestyles is as deeply and uniquely embedded in the individual as the more universally accepted differences in academic, athletic, and musical talents. People tend to believe attitudes can be "willed" even if aptitudes cannot. We are not so sure. Changing someone's predisposition toward trying hard or toward illegal immigrants often is no easier than altering that person's musical ability, preferred writing hand, or proficiency at commutative algebra.

The bottom line is that innate aptitude is unlikely as trivial a factor as Gladwell implies. Most of us could not skate like Wayne Gretzky, play guitar like George Harrison, sing like Adele, think like Stephen Hawking, dribble or shoot like Lionel Messi, or jump like Michael Jordan no matter how much we practiced and no matter where on the calendar our birthdays happened to fall. But it is arguably a bigger mistake to believe that innate attitudes do not exist and therefore that all people are dispositionally the same when it comes to work ethic, favored recreational and occupational pursuits, or even preferences for the best way to organize and run mass-scale society. Most people, including Gladwell, accept that individuals vary somewhat in aptitudes, but most people tend to be less willing to accept that differences in attitudes are shaped by similar sorts of forces (things like biological and cognitive dispositions), yet they are.

It is a significant adjustment to think of attitudes as products not just of our environments or situations but also of biologically based predispositions; yet attitudes are undeniably based in what people think and feel, and thinking and feeling are undeniably physical processes. It is possible to tell if people view images as stimulating merely by looking at patterns of their physiology. If an electrode is placed in the brain's motor cortex, it can pick up the physical signals of a paralyzed person's thoughts and translate those signals into physical action such as raising an arm. Neuroscientists are even getting to the point where they can tell which movie stars people are thinking of (Zack Galifianakis or Will Ferrell in one study) by looking at images of their brains.[29] The bottom line is that physical processes are connected with thinking and feeling. From here it is a pretty small step to accept that these physical processes might vary in meaningful ways from one person to another. In short, it is not just variation in situations that matter to behavior, but variation in the physical processes that predispose people to have different thoughts and feelings, and thus behavioral responses, to exactly the same situation or environmental stimulus.[30] That is potentially important because as we are about to demonstrate, many of those variations have big implications.

Abby Normal

The *Diagnostic and Statistical Manual of Mental Disorders* (DSM) is the Bible of clinical psychiatry and psychology and its guidelines have enormous social impact. The DSM classifies mental disorders by describing and listing the symptoms associated with all common, and many not so common, psychological maladies. In a very real sense, you are not considered depressed, schizophrenic, or autistic until the DSM says you are. This is because institutions like insurance companies, schools, and social service organizations require a diagnosis before agreeing to reimbursement, medication, or special tutoring, and the DSM is the accepted standard for making and justifying that diagnosis.

The DSM has been used for this purpose for decades; the original DSM-I was published in 1952 and has gone through four revisions since then. Until recently, the basic approach to diagnosis went something like this: The DSM offered a laundry list of symptoms associated with a disorder, and an attending professional tested for those symptoms. If those tests were positive for the specified number of symptoms, you were diagnosed with the disorder. If not, you were not diagnosed with the disorder. It was an in-or-out sort of deal.

Consider the DSM-IV's "Diagnostic Criteria for Autistic Disorder," which listed 12 symptoms, split equally into three distinct categories: impairments in social interaction, impairments in communication, and repetitive and stereotyped patterns of behavior. A diagnosis of autism required the presence of at least 6 of the 12 symptoms, with at least two of these from the impaired social interaction list, and at least one each from impaired communication and repetitive behaviors. Abnormal functioning also had to be detectable prior to age 3. People who met these criteria were diagnosed as autistic. People who had five symptoms, or who had six symptoms not distributed appropriately, were not diagnosed as autistic, though patients failing to meet the Autistic Disorder criteria might meet the Asperger's Disorder criteria. Asperger's had basically the same list of symptoms, but not as many were required for a diagnosis and the "by age 3" stipulation was absent. Again, though, according to the DSM-IV guidelines, patients were diagnosed dichotomously: Patients did or did not have Asperger's Disorder. This approach to psychological conditions divides the world into those with and without psychological pathologies—there's "normal" and "abnormal" and no other variation.

This all seems more than a bit arbitrary. How can we be sure that the proper cutoff point for Autistic Disorder is six symptoms rather than five? How do we know that the "two from column A; one from columns B and C" approach is the appropriate one? Are there really 12 symptoms of autism? Maybe we missed one, or maybe two of those listed are really the same thing. Objectively establishing verifiable criteria to divide everyone into they-have-it-or-they-don't categories is so difficult as to be impossible.

For purposes of insurance claims and special education labels, putting people into the "diagnosed" or the "undiagnosed" category makes a certain amount of practical sense, but for those of us trying to understand actual individual differences, it makes no sense at all. Psychological differences, be they reflected in social skills or political orientations, are differences of degree rather than kind, and dichotomization of the population covers up a great deal of meaningful variation. The current guardians of the DSM seem to agree and the recently released DSM-5 (no more Roman numerals) contains an important conceptual shift acknowledging that there are various shades of dysfunction. Mental disorders are no longer viewed as discrete conditions but as the extremes of a spectrum. Autism, Asperger's, and Childhood Disintegrative Disorder have been collapsed into a single category (much to the dismay of those previously diagnosed as Asperger's) called Autism Spectrum Disorder (ASD), but instead of individuals merely being labeled ASD or not, they are also given a rating of the severity of their disorder.

This important change in clinical psychology is very much in line with our thinking. We only differ in the sense that the basic concept should not be limited to those who have qualified as having a disorder. Individuals diagnosed with ASD differ from each other, but individuals who have *not* been diagnosed with ASD also differ from each other in terms of the same sorts of underlying psychological dispositions. Does anyone seriously believe that once those individuals diagnosed with Autism Spectrum Disorder are set aside, the rest of the population is exactly the same with regard to social skills and proclivity for repetitive behaviors? As academics, we interact with people from a variety of disciplines, including engineering and literary studies. This interaction makes us highly skeptical that all people *not* clinically diagnosed with ASD are dispositionally identical in, say, their ability to discern the emotions of others, in their comfort level with anecdotes as opposed to analytical systems, and in the degree to which they "find social situations easy" or "are good diplomats." People with a tendency for longwinded, one-sided conversations, repetitive behaviors, or a dearth of eye contact might not qualify as having ASD. They might just be socially awkward blowhards—some of whom managed to become professors.

The point is that a failure to be diagnosed with any mental malady does not automatically qualify a person for the label "normal." Our view is that there is a continuum for pretty much any psychological function or orientation. At some point on that continuum a diagnosis might be justified, but that does not mean variation ceases and collapses into a single "normal" category below that threshold. Autism researcher Simon Baron-Cohen noted this situation long ago. He developed collections of survey items that assessed location on a systemizing scale (sample item: "When I look at a building, I am curious about the precise way it was constructed"), an empathizing scale (sample item: "I am good at predicting how someone will feel"), and something he called an autism spectrum quotient (sample item: "I prefer to do things the same way over and over again"). Each of the three collections of survey items has at least 50 items and on each item it is possible to strongly agree (3), slightly agree (2), slightly disagree (1), or strongly disagree (0), so theoretically people could score anywhere from 0 (0×50) to 150 (3×50). In other words, this is a scale that captures a wide range of variation in terms of the degree to which someone—anyone—is an empathetic or a systemizer.[31]

Just because two people do not qualify for a disorder does not mean that they are identical or undeserving of our attention. If the full range of variation were appreciated, the conclusion would be (and is in the neuroscience community) that there is no neurotypical, there is no normal. The contrast between the old and new visions of variation is depicted in Figure 3.1, using autism and systematic thinking as the example. Widespread recognition of the full range of behaviors would permit a healthier and more accurate way of looking at the human condition than forcing everyone into restrictive, two-category solutions such as normal/abnormal or autistic/not autistic. The unsubstantiated assumption that all people who fail to merit an official DSM diagnosis

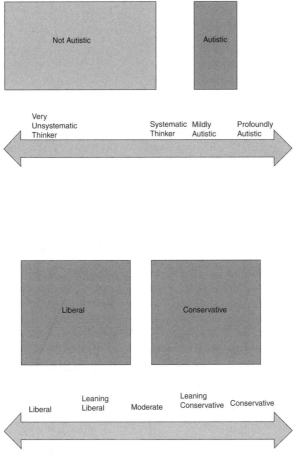

Figure 3.1   Shifting from Discrete Categories to Continuums

are physiocognitively the same creates problems. For starters, it increases the stigma placed on those who happen to fall, perhaps barely, on the "disorder" side of one of the very arbitrary lines that have been drawn. If the tremendous variety of behaviorally relevant predispositions across people were recognized, being labeled with a disorder would be less of a big deal. Some people think more systematically than others, sometimes to the point of frustrating those around them but not to the point of being diagnosed as autistic. We all occupy unique locations on each behavioral spectrum. It is time to give more than lip service to diversity; it is time to acknowledge that diversity extends all the

way to behaviorally relevant biological differences. The concept of normality
needs to be ditched; however, doing so will not be easy and will have big and
perhaps scary social implications. Take, for example, the implications for the
legal system.

Telling Right from Wrong

Daniel M'Naughton was born a bastard in 1813 in Glasgow and suffered the
indignities and financial challenges commonly associated with being illegiti-
mate in that era. His mother died and young M'Naughton became an apprentice
woodworker to his father. When it became clear he would never be made a
partner, he left, trying his hand at acting before setting up his own woodturning
shop in 1835. Outside of an occasional foray into radical politics, he seemed a
good citizen; his business was successful and he was considered frugal, industrious, sober, and intellectually curious.

Then he went off the deep end.

In 1840 he sold his shop and started spending a good deal of his time
in London, though no one was quite sure what he did there. He began
complaining—first to his father, then to the local police, and finally to an MP
(Member of Parliament)—that Tory politicians were persecuting and spying on
him, all part of a plot by Prime Minister Robert Peel to ruin him. On January 20,
1843, M'Naughton, apparently under the impression that a pedestrian walking
along Downing Street was Peel, snuck up behind and shot him at point blank
range. M'Naughton was quickly nabbed by the police. His victim was in actuality Edward Drummond, Peel's personal secretary; Drummond died five days
later of complications from his wounds.

At his trial, defense and prosecution agreed that M'Naughton was seriously
delusional. Medical witnesses testified to his condition and acquaintances
from Glasgow gave accounts of his increasingly bizarre behavior. The upshot
was that the jury sided with the defense and, rather than being sent to the
gallows, M'Naughton was delivered to Bethlem Hospital's wing for "Insane
Persons Charged with Offenses." Outrage ensued. Popular sentiment was that
M'Naughton deserved to hang. Queen Victoria was not amused. The victim
of more than one assassination attempt herself, she let it be known that she
was deeply bothered by the verdict, and her entreaties motivated the House

of Lords to insist that the 12 judges on the "Court of Common Pleas" answer several questions on the theory and practice of the insanity defense. The judges complied with a set of points that still shapes thinking on legal culpability in many parts of the world today.[32]

The essence of what came to be known as the M'Naughton Rules is that an individual should not be held fully legally responsible if "laboring under such a defect of reason from disease of the mind as not to know . . . that what he was doing was wrong." Thus, the ability to tell right from wrong became the key test for claiming insanity, which is accepted as a legitimate plea in most western countries and in all but four of the United States. The M'Naughton Rules' assertion that the ability to tell right from wrong is the key to determining whether an individual is or is not legally culpable endures as a fundamental legal standard.

The distinction also endures as completely impossible, wholly artificial, and embarrassingly inappropriate. Sorting the world into those who can and those who cannot tell right from wrong is as preposterous as the notion that people are either autistic or not. Psychologists develop tests, defendants develop acting skills, and dueling expert psychological witnesses ("the defendant could not tell right from wrong at the time of the crime"; "oh yes he could") develop fat bank accounts. In reality, the courts have little idea how to distinguish those who have a "disease of the mind" from those who do not. The official standard for intellectual disability is "significant limitations both in intellectual functioning and in adaptive behavior." Given the amorphous nature of this phrase, courts often fall back on arbitrary age or IQ cutoff points.

In the United States, the Supreme Court has ruled the death penalty cannot be applied to people younger than 17. Those 18 and older with an IQ of 70 or less also tend to be exempted from capital punishment because they are frequently considered intellectually disabled. Is there really a life and death difference between IQ measures of 71 and 69? Are we that confident in our ability to measure intellectual ability? While we are at it, what magical event transpires on eighteenth birthdays to make people no longer intellectually disabled? Is it not likely that some people mature, neurologically and otherwise, later than others? The eagerness to draw lines and divide people into discrete groups creates ridiculous distinctions and situations, such as defendants suddenly developing a keen interest in convincing jurors that they are mad as hatters.

In truth, some people know right from wrong, some do not, and some sort of know but are fuzzy about the distinction.

The legal issue of automatism (an absence of voluntariness or culpability) is even more interesting; it raises the possibility that some people know right from wrong perfectly well but cannot stop themselves from doing wrong. Take Amanda Clarke, an apparently law-abiding citizen caught shoplifting. She put a large number of items in her shopping cart, paid for them and left the store. She also put butter, coffee, and mincemeat in her purse, left without paying for them, and was flabbergasted when security personnel found them there. She had no idea how those items materialized in her bag. She was charged anyway. At trial, the defense pointed out that Clarke, then 58, was diabetic and chronically absent-minded, and suffered from severe depression. How could she be held responsible for something that she did not consciously do? The court would have none of it and declared that short periods of absent-mindedness did not amount to a "defect of reason" or a "disease of the mind."

Of course, this distinction immediately raises the question of what exactly qualifies as a disease of the mind, and the courts have made a complete hash of clarifying that point. Amanda Clark's chronic depression apparently is not enough. Crimes committed while sleepwalking, though, are occasionally viewed as instances of automatism. It is easy to see that someone in the midst of a grand mal seizure should not be held culpable, but what about seizures that are not quite so grand or not quite so mal? In the United Kingdom, diabetics have been permitted to use hyperglycemia but not hypoglycemia in their defense. People who voluntarily place themselves under the influence of alcohol or mind-altering drugs raise another set of issues. Are they fully responsible because they purposely diminished their ability to make good decisions and do right rather than wrong? What about someone who was slipped a mickey without his or her knowledge and then committed a heinous act? Is a chronic addict different than an occasional user?

Consider the case of a 40-year-old, happily married man (name understandably withheld in the official reports) who suddenly began to show an intense interest in child pornography. Soon, he was spending virtually all his time looking at pornographic websites and literature and soliciting young women. He also suffered from increasingly painful headaches. His wife persuaded him to seek medical counsel and a neurologist found a very large tumor in the area of the orbitofrontal cortex (OFC). Surgery was performed and, with the tumor removed, his interest in child pornography evaporated . . . until several months later, when it came back. A return visit to the neurologist revealed part of the tumor had been missed and was growing. One more surgery forever removed his fixation with child porn. This unfortunate fellow had clearly violated the law and knew that what he was doing was wrong on a number of levels but was utterly powerless, as each of us would be, to counteract physical pressure on the OFC. What should the courts do with him?[33]

The case of tumor-induced child porn addiction is a clear illustration that distinguishing those who do and those who do not know right from wrong is not nearly enough. Some people know right from wrong but, perhaps because of the constitution of their frontal cortex, find it impossible to "do right."[34] Accepting that major biological variations affect behavior makes it impossible to deny that minor variations do as well. Even without a tumor pressing on their orbitofrontal cortex, individuals have varying densities of chemical receptors at key areas in the brain, differently shaped neural organs, and neurotransmitter levels in synapses that are highly variable. The effectiveness of drugs such as Ritalin and Prozac makes it clear that decisions and behaviors are biological. If artificially adjusting chemical levels in the brain affects attitudes and actions, naturally occurring variations would have the same effect. Still, the courts do not recognize such variations. Just as laziness must be the cause of not working hard, a criminal lack of discipline must be the reason someone who is mentally capable of discerning right from wrong would not do right. Such thinking ignores the growing neurological evidence that some people, for reasons not fully under their control, have to struggle very hard to do what is right or what is sensible even though they do not qualify for the label "intellectually disabled."

This was demonstrated in a famous study by neuroscientist Antonio Damasio. He used cards that when turned over reveal a payoff amount—sometimes positive, sometimes not. The cards were arranged into different decks. Some decks led to good overall outcomes (meaning if you turned over all the cards in that deck you would win money); other decks were less favorable (turning over all cards in the deck would result in a net loss). The "bad" decks, however, did have individual cards with really big payoffs, and individuals with certain conditions, such as lesions in the ventromedial prefrontal cortex, had difficulty refraining from playing the bad decks even when they knew they would end up being worse off.[35] The lure of the occasional big payoff apparently overpowered them. Knowing what should be done is not tantamount to doing it. Predispositions affect both knowledge and action.

The only way for society to function may be with a legal system that, except in the most egregious cases, denies it is biologically more challenging for some people "to do good" and that asserts that all nonclinical people are the same in terms of their ability to know right from wrong. This, however, does not mean we need to convince ourselves that they actually are. Pretending that all people have identical behaviorally relevant biological dispositions is intellectually dishonest and contradicts much empirical evidence. Our purpose here is not to argue for excuse making and leniency. Rather, our point is that people vary in ways that defy dichotomous categorization. Even though the evidence is overwhelming that people have all sorts of predispositions and that these predispositions vary from one person to another in subtle ways that make each of us unique, this evidence is seldom taken seriously. In fact, discussions concerning behaviorally relevant biological differences are rare—with one big exception.

Pro-Choice?

"The opposite of homosexuality is not heterosexuality, it is holiness." So sayeth Alan Chambers, leader of Exodus International, an organization discouraging "same-sex attraction" by hosting Christian-themed reparative therapy workshops ("boot camps" to detractors). The people at Exodus International attribute same-sex attractions to abuse, shame, self-hatred, and unhealthy relationships with family and peers. Since environmental forces typically create same-sex attractions, it makes sense to Exodus International that environmental forces be employed to neutralize them. Thus, the week-long workshops promise "transformations and healings," and include sessions such as "Overcoming Guilt and Shame" and "Walking Away from the Lesbian Mentality."

Unlike most organizations that host camps of this ilk, Exodus International does not insist that same-sex attractions are optional and the organization is open to the possibility that long-term sexual predispositions exist. Teachers at its events openly admit that they continue to be afflicted with same-sex attraction and that mitigating rather than eliminating same-sex feelings is the more realistic goal. After acknowledging her same-sex attraction, one instructor proudly proclaimed that, through counseling and force of will, it is like "elevator music to me now." Chambers himself echoes this philosophy, saying, "I didn't choose my same-sex feelings but I do choose how I'm going to steward them." He continues, "I lead a life of denial and I love it."[36]

Other organizations and individuals go further and maintain that same-sex attraction has no deeper basis and can be eliminated if the individual is willing to make a different choice. Most gay and lesbian people and their supporters, on the other hand, believe sexual orientation is much deeper than an environmentally shaped choice. They believe biological predispositions render individuals unable to choose sexual orientation. Many gay people realized from a surprisingly early age that they differed from the heterosexual majority and are dumbfounded that sexual orientation could be viewed by anyone as a choice. This debate over the source of variation in sexual orientation is the closest modern society comes to addressing behaviorally relevant biological differences. As such, it serves as a template for pondering the nature of behavioral predispositions in realms other than sex.

Is sexual orientation a biological predisposition or is it just a choice that is made, rather like choosing which car to buy? People have predispositions in all areas of life, from personality to occupation and from politics to leisure pursuits. Given this, it should not be surprising to find predispositions regarding sexual partners. Such predispositions make it difficult to casually trade one sexual orientation for another. As with all the other areas we have been discussing, however, one of the major barriers to rational thought is the desire to dichotomize. The truth is not everyone is either gay or straight. Some only lean gay; some only lean straight; some are bisexual; some are asexual; and some have preferences that cannot be described in a PG-13 popular science book. People not as deeply predisposed in one sexual direction or the other probably could be influenced by their environment, but just because some people's orientations are plastic does not mean everyone's are. Efforts at conversion via boot camp have resulted in formerly gay people operating as heterosexuals, but these occasional "conversions" should not be taken as evidence that everyone is equally convertible. For many, no amount of environmental manipulation is going to change their sexual orientation.

Alan Chambers' program tamps down the same-sex attraction of some people, while others are completely oblivious to even the most intrusive ministrations. They can wrestle with their orientations, pray that they change, and spend all their waking hours in workshops, but they will still be unabashedly gay. Behaviorally relevant biological predispositions exist, but they do so in varying degrees and therefore are not determinative. Whether the issue is politics or sex, probabilistic thinking is crucial; otherwise, debate will be characterized by a profitless trading of anecdotes and little progress toward understanding.

Conclusion: The Politics of Difference

People's differences run deep. We are not all born with the same "slates." We come into the world with much on our slate and the environment we encounter piles on its idiosyncratic touches. Though the prevailing view both in academic and folk wisdom is that this individuation is not particularly enduring or biological, in more self-reflective moments most people accept that they have longstanding biases and predispositions. Our claim is that these predispositions are biologically measurable and connect to a variety of generic psychological and cognitive patterns.

Accepting such dispositional differences calls into question the assumption that down deep people are really all the same, except for those who suffered some trauma or malady that has left them abnormal, disordered, or unable to tell right from wrong. Dispositional differences suggest that the standard academic practice of exposing two groups of randomly assigned people to different stimuli, computing the average difference in behavior between the groups, and then declaring that the situation causes people to be generous or subservient to authority figures misses a critically important part of the story: the remarkable variation that exists around those averages. Dispositional differences suggest that behaviorists, evolutionary psychologists, classical microeconomists, experimental social psychologists, political theorists, Communists, social engineers, popular commentators, standard social scientists, legal authorities, diagnosticians, and fans of the noble savage theory all miss the same important part of the story.

Are political orientations immune from the shaping influence of deepseated, behaviorally relevant biological predispositions? In point of fact, given the emotionality suffusing it, politics more than most elements of life is shaped by individual predispositions—or at least it can be. Parallel to the role of biology in the sexual preferences of gay and straight people, the political preferences of some liberals and conservatives connect much more to biological predispositions than is the case for others. And of course, this predispositional approach also helps to account for the fact that not all people fit neatly into just two camps. Just as some people are bisexual and some asexual, some are politically moderate and some are apolitical. Not all liberals are created equal and neither are all conservatives.

Politics addresses fundamental issues common to all mass-scale social systems and is central to the modern human condition. People often have unique behaviorally relevant predispositions that affect their attitudes and actions in all areas of life, including their preferences for the appropriate solutions to political problems. These politically relevant predispositions can be glimpsed on four separate but interlocking levels: psychological orientations and tastes, patterns of cognition, physiological responses, and genetics, and in the next four chapters we explore the empirical evidence at each of these levels, suggesting in the end that political orientations are indeed grounded in physiopsychological predispositions. We begin by placing political tastes in the context of the larger package of tastes for art, food, humor, social situations, and psychological arrangements.

Notes

1   Milgram, "Obedience to Authority."
2   Zimbardo, *The Lucifer Effect: Understanding How Good People Turn Evil.*
3   Henrich et al., "In Search of Homo Economicus."
4   Cameron, "Raising the Stakes in the Ultimatum Game: Experimental Evidence from Indonesia"; and Andersen et al., "Stakes Matter in Ultimatum Games."
5   Hoffman and Spitzer, "Entitlements, Rights, and Fairness: An Experimental Examination of Subjects' Concepts of Distributive Justice."
6   Solnick and Schweitzer, "The Influence of Physical Attractiveness and Gender on Ultimatum Game Decisions."
7   Bohnet and Frey, "Social Distance and Other-Regarding Behavior in Dictator Games: Comment"; and Charness and Gneezy, "What's in a Name? Anonymity and Social Distance in Dictator and Ultimatum Games."
8   Milgram, "Obedience to Authority," 205.
9   Sanfey, "The Neural Basis of Economic Decision-Making in the Ultimatum Game"; and de Quervain et al., The Neural Basis of Altruistic Punishment."
10 McCabe et al., "A Functional Imaging Study of Cooperation in Two-Person Reciprocal Exchange."

[11] Vickery and Chun, "Object-Based Warping: An Illusory Distortion of Space within Objects."

[12] Appropriately, one of Skinner's best-selling books was called "Beyond Freedom and Dignity."

[13] Watson, "Experimental Studies on the Growth of the Emotions," 82.

[14] Harlow et al., "Total Social Isolation in Monkeys."

[15] Garcia et al., "Conditioned Aversion to Saccharin Resulting from Exposure to Gamma Radiation."

[16] Cook and Mineka, "Observational Conditioning of Fear to Fear-Relevant Versus Fear-Irrelevant Stimuli in Rhesus Monkeys."

[17] Seligman, "Phobias and Preparedness."

[18] Turcsan et al., "Trainability and Boldness Traits Differ between Dog Breed Clusters."

[19] Cosmides and Tooby, "What Is Evolutionary Psychology?"

[20] Maestripieri et al., "Father Absence, Menarche, and Interest in Infants among Adolescent Girls."

[21] Tooby and Cosmides, "The Past Explains the Present: Emotional Adaptations and the Structure of Ancestral Environments." For a critique of this view, see Cochran and Harpending, *The 10,000 Year Explosion*.

[22] Camerer et al., "Neuroeconomics: How Neuroscience Can Inform Economics."

[23] Mueller, *The Public Choice Approach to Politics*.

[24] Thomas Hobbes, *Leviathan*.

[25] Locke, *Two Treatises on Government*.

[26] Durkheim, *Division of Labor in Society*.

[27] Freeman, *Margaret Mead and Samoa: The Making and Unmaking of an Anthropological Myth*.

[28] Gladwell, *Outliers*.

[29] Suthana and Fried, "Percepts to Recollections: Insights from Single Neuron Recordings in the Human Brain."

[30] We have overstated the lack of attention to individual differences to make this point but not by much. The great majority of social science work; of political theorizing; and of popular, wishful writing, as well as of the mega-academic movements of behaviorism, evolutionary psychology, and classical microeconomics all stress the universality of human nature and have precious little to say about individual differences. Still, we should acknowledge that a pocket of scholars has been taking individual differences seriously for quite some time. This interest transcends disciplines and even has its own professional organization, the International Society for the Study of Individual Differences, and its own academic journal, *Personality and Individual Differences*. These sorts of academic pockets, though, are still more exception than rule, which is a pity because we are living in a time when the technology and techniques available—brain scans, gene sequencing, hormonal assays, physiological and cognitive measurement—provide the tools to make this a golden age of studying individual-level differences.

[31] Baron-Cohen, *The Essential Difference: Men, Women, and the Extreme Male Brain*.

[32] Moran, *Knowing Right from Wrong: The Insanity Defense of Daniel McNaughtan*.

[33] Burns and Swerdlow, "Right Orbitofrontal Tumor with Pedophilia Symptom and Constructional Apraxia Sign."

[34] Eagleman, *Incognito: The Secret Lives of the Brain*.

[35] Damasio, "The Somatic Marker Hypothesis and the Possible Functions of the Prefrontal Cortex [and Discussion]."

[36] Bannerman, "The Camp That 'Cures' Homosexuality."

C H A P T E R **4**

Drunk Flies and Salad Greens

You're either a liberal or a conservative if you have an IQ above a toaster.

Ann Coulter

Political issues are decided at the table.

Lucien Tendret

One of the great puzzles of biology is why some insects like beer. For example, many fruit flies love the stuff. Why would fruit flies like beer? Pretty much anyone who has ever swigged something sweet in the summer heat knows that flies have a taste for sugary liquids. Fruit flies primarily feed on decaying fruit and, as they cannot chew, they slurp up nutrients through a proboscis, a process roughly equivalent to drinking fruit juice through a straw. Fruit juice has a high sugar content, and sweets like sugary water signal an all-you-can-eat-buffet to the average fruit fly. Beer, though, is not sweet. Brewing beer involves a fermentation process where yeast breaks down sugar into ethanol and carbon dioxide. After this, only trace amounts of sugar remain in beer, so whatever draws flies to the drink, it is clearly not the residual sugar floating around in the average pint. So why would an organism whose survival basically depends on sucking down sugary liquids be so fond of beer?

A group of scientists at the University of California, Riverside (UCR) hypothesized that the attraction might actually be glycerol, a substance present in concentrations of about 1 percent in beer (roughly 1 to 2 grams per liter). They had a grad student, Zev Wisotsky, pick up some beer at the grocery store—a pale ale to be precise, a type of beer specifically chosen for its low sugar and high glycerol content. In a controlled experiment, Wisotsky basically offered the flies a choice between beer or sugar water. Though some did not, many flies went for the beer presumably because even though glycerol is not a sugar it tastes sweet.[1] The UCR scientists were not done, however. They went a step further and through a process of genetic modification were able to breed flies that either liked sugar water more than beer or beer more than sugar water, thereby demonstrating that the variation in wild fruit fly beverage preferences is attributable, at least in part, to genetics.

When it comes to tastes and preferences, humans are not that different from fruit flies. For a start, there is great variation from organism to organism. Just as some flies like sugar water while others prefer beer, some people prefer Kool-Aid and some beer. Some people like savory foods while others have a sweet tooth, and the textures and aromas that are delightful to some are repulsive to others. These variations in tastes and preferences are not limited to food. People discriminate in everything from music to fashion, art to humor, and even styles of storytelling. Not surprisingly, people move toward those stimuli they find tasty, pleasing, satisfying, or funny and avoid those stimuli they believe to be disgusting, disturbing, or just plain boring. It is important to note that the same stimuli can be pleasing to one person and not so pleasing to another, and these variations in tastes and preferences lead to individual differences in attitudes and behavior. At the same party, some will like the music choices; others will not. Some will munch peanuts; others head for the veggie tray. Some will find the host's anecdotes amusing; others will find them in poor taste. Some will genuinely admire the guest wearing the latest fashion; others will think he'll live to regret appearing in public in something so outrageous. Some will throw themselves eagerly into the novelty of a party game; others will hang back out of concern that they will make fools of themselves in public. Some will be teetotalers; others tipplers.

All this is relevant to the theme of this book because variation in tastes and preferences, and more broadly in the personality tendencies and values that shape what we find pleasing or annoying, is connected to political orientations. In this chapter we will describe some of the different tastes and preferences that distinguish conservatives and liberals and take a critical look at explanations of the source of those differences and the reasons they relate to politics. Tastes and preferences cover a good deal of diverse ground and so will we in this chapter. Where we are headed is a basic theory of why people are so different, not just in their politics but in their broader subjective tastes and preferences. Just as flies' taste for beer is biologically based and relates to their behavior, humans' tastes are often biologically based and relate to their behavior, right down to political orientations.

Meatloaf Conservatives and White Wine Liberals

One of the issues that divided the 2008 Republican and Democratic U.S. presidential candidates was their differing perspectives on aromatic, peppery-tasting salad greens. The big veggie controversy was rooted in a comment made by Democratic nominee Barack Obama during his primary campaign in Iowa. In making a point about farmers not seeing more income in their pocket despite price increases in grocery stores, he said, "Anybody gone into Whole Foods lately and seen what they charge for arugula?"[2] This comment came back to haunt him as his opponent in the general election, Republican John McCain, used it to highlight the clear choice that voters faced. One candidate was a wine-sippin', Ivy League–educated liberal with a taste for fancy overpriced lettuce. The other candidate was an all-American guy who preferred regular food like red meat and gravy (McCain professed not to be a big veggie man).

This may all sound like caricature: food snobs who lean left when passing on their policy opinions, not just their recipes; meat-and-potatoes conservatives who have a taste for, wouldn't you know, meat and potatoes. There is something vaguely Monty Python-esque in the notion of the two candidates for the most powerful executive office on the planet appealing to voters on the basis of their favorite foods. Still, McCain was onto something. Political differences are not just aired around the dinner table; they have a strong relationship to what we like to see on the dinner table.

A couple of the largest studies to provide such evidence were carried out by the website Hunch.com. Hunch describes itself as a means to "personalize the

Internet by helping you to share and discover great recommendations about all sorts of topics." In essence, it's a collective-intelligence decision-making tool— a recommendation engine that utilizes the inputs of millions of users to refer people to stuff they will like. It works by getting users to answer numerous questions about themselves and then looking for patterns in the responses. For example, Hunch at one time uncovered that if you were an iPhone user you were also more likely to say you enjoy Rice Crispies, BBC television shows, and (not surprisingly) Macs over PCs. On the other hand, if you were an Android phone user you were more likely to prefer Brooks Brothers clothes, Honey Nut Cheerios, and PCs over Macs.

Among the questions Hunch asks its users is whether they tend to support liberal or conservative politicians. In 2009 Hunch did an analysis of food preferences based on 64,000 users who indicated a political preference. One of their findings provides empirical evidence for the salad spat between McCain and Obama: Liberals picked arugula as their choice of salad green more than twice as often as conservatives. The study was updated a few years later based on 400,000 responses to the political question, and again consistent differences in food preferences were found between liberals and conservatives.[3] Care has to be taken in drawing conclusions too firmly from these sorts of studies because they are not based on random samples. In other words, we can only be sure that the results apply to people who use Hunch and who answered the question on political preferences. As a group, that set of users is not particularly representative of the U.S. adult population; for one thing, they tend to skew liberal. Still, even with such limitations, we can say that among at least one group consisting of hundreds of thousands of people liberals and conservatives possess distinct culinary preferences. Moreover, the fact that these results mesh so nicely with those derived from representative samples (see below) is comforting.

The netizens of Hunch are not the only ones to reveal these differences. Neuropolitics.org is a website devoted to reporting on the differences between liberals and conservatives and it periodically surveys its readers on a wide variety of topics, including food preferences. The site reports findings similar to Hunch's: Conservatives like beef and liberals are more likely to be vegetarians. Conservatives also report weighing more, which is not surprising given that they don't exactly seem to be eating lightly.[4] These results have the same sort of limitations as the Hunch study, but they at least show that the basic liberal-conservative difference in food preferences is replicable.

The difference is also supported by mounds of anecdotal evidence. After all, McCain was far from the first actual or aspiring Republican White House occupant to turn up his nose at vegetables. President George H. W. Bush famously didn't like his broccoli and his son George W. Bush liked plain pretzels to snack on when he was president and choked on one while watching TV at the White House in 2002. Turning to Democrats, rather than highlighting processed, high-carb, sodium-laden snack foods like pretzels, the Obamas went organic and started their own kitchen garden in back of the White House.[5]

These examples all seem to support a general pattern: When it comes to food, liberals are more likely to seek out the new, the novel, and the exotic while conservatives are more likely to stick with the tried and true. Since the evidence was a bit unsystematic, however, we wanted to test this very proposition in a study of our own. We took a random sample of about 350 adults residing near a particular city in the Midwestern United States and gave them a lengthy survey

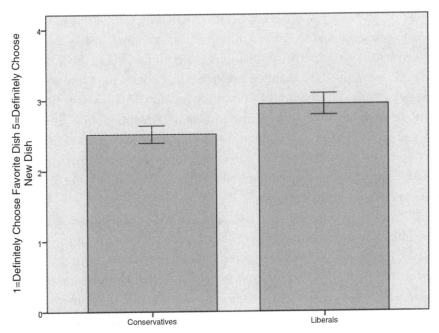

Figure 4.1   Ideology and Food Preferences

asking all sorts of questions including items on their tastes and preferences. One of the questions was: "Given a choice between a favorite meal and a new and exotic dish you've never tried before, which would you choose?" The response options were in the form of a 5-point scale where 1 corresponded to "definitely choose the favorite dish" and 5 corresponded to "definitely choose the new dish." Self-identified conservatives averaged a 2.5 on this scale, while liberals averaged about 3.0. (See figure 4.1, page 93.) That may not seem like a lot, but the difference was statistically significant. In other words, it supports the general pattern and provides systemic evidence that the informal surveys as well as the frequently repeated anecdotes are based on a real relationship.

If nonpolitical differences between liberals and conservatives were limited to food, they might amount to nothing more than an amusing piece of trivia. Yet what is really interesting is that this general pattern seems to extend well beyond food preferences. We asked our survey respondents other questions about their likes and dislikes. What did they find funny? What sort of fiction did they like to read? What sort of art did they like to look at? How into new music were they? Sure enough, on the majority of these items we found statistically significant differences between liberals and conservatives. Across a range of topics, the mean responses of liberals consistently favored the new experience, the abstract, and the nonconforming. Conservatives just as consistently favored traditional experiences that were closer to reality and predictable patterns. Conservatives, for example, preferred their poems to rhyme and fiction that ended with a clear resolution. Liberals were more likely to write fiction and paint, or attend a music concert. Experimental, arrhythmic verse, amorphous story lines, and ambiguous endings just do not trip the triggers of many conservatives and, perhaps relatedly, they are less likely to be performers, a fact that is all too apparent from the announced political affiliations of comedians, rock stars, and Hollywood actors.

Differences in art preference also are clear, with conservatives more likely than liberals to avert their eyes from colorful, abstract work in order to look at realistic landscapes. A study by psychologist Glenn Wilson from decades ago mirrors our findings on art preferences by finding that abstract, complex art is less appreciated by conservatives.[6] Other research confirms that conservatives are less likely to be involved in artistic activities and there is at least one study besides ours reporting a systematic liberal-conservative divide on poetry preferences.[7]

Studies even have uncovered liberal-conservative differences in humor preferences. For example, Wilson finds that jokes that "failed to provide resolution of incongruous elements" are less likely to hit the conservative funny bone.[8] The distinction appears to be stable across cultures, or at least Wilson found it in the United Kingdom in 1969 and we found a similar pattern in the United States in 2010. Liberals seemed more amused by, "I almost had a psychic girlfriend but she left me before we met," or "I planted some bird seed . . . a bird came up . . . now I don't know what to feed it," while conservatives preferred puns or jokes such as, "You know you're in trouble when at the control tower there's a note taped on the door that says 'back in 5 minutes.'"

One of the most fascinating recent studies suggesting that liberals and conservatives differ on much more than politics was conducted by Dana Carney of Columbia University and several colleagues. With permission, they systematically inventoried the contents of bedrooms and office spaces of roughly 150 people. They found that tastes and preferences not only correlated with political orientation, but were manifested in people's personal living spaces. For example, conservatives were more likely to have items associated with organization and neatness, such as laundry baskets, postage stamps, and event calendars, while liberals were more likely to have art supplies, stationary, and a broad variety of music CDs. Carney's wide-ranging study concluded that political orientation seems to reflect everything from behavioral patterns to travel choices to the way we "decorate our walls, clean our bodies and our homes, and . . . choose to spend our free time."[9] Other studies show that particular leisure pursuits (soccer vs. NASCAR) and career paths are more attractive to liberals than conservatives and vice versa. Academics, for example, are well known to be a left-leaning lot.[10]

And it is not just social scientists and websites that have picked up on the fact that liberals and conservatives differ in their tastes and preferences well beyond politics. The Republican National Committee hired market research firms to analyze partisan consumption patterns so they could better target political messages. Among their findings: A partisan divide is clearly evident in car ownership.[11] At the high end, Republicans tend to favor Porsches (nearly 60 percent of Porsche owners identify as Republicans), while Democrats favor Volvos. At the lower end, Republicans tend to like American-made cars; Democrats prefer Hyundais. Republicans tend to show more loyalty to a particular car brand while Democrats shop around more. In other words, Republicans seem to favor established, traditional automobile manufacturers and stick with them. Democrats have weaker brand loyalty and are more willing to check out alternatives.

The Center for Responsive Politics looked at the stock investments of members of Congress and found a clear divide on investment patterns. Republicans tend to favor industrial and resource-extraction stocks (think BP and Exxon); Democrats prefer high tech stocks.[12] In other words, Democrats favor the stocks of companies dependent upon creativity and new thinking while Republicans tend to favor companies that deal in something tangible. Other factors, such as the source of campaign contributions, likely factor into the investment decisions of politicians, but these findings fit with the overarching story.

So it is not just preferences for food that differentiate conservatives and liberals; it is a large set of preferences regarding the experiences that bring satisfaction or frustration, interest or boredom, pain or pleasure. These sorts of differences should not be exaggerated (think probabilistically). Some arugula-loving novelists who like going to rock concerts in their Hyundais lean politically to the right, just as some burger-eating, Porsche-driving, poetry-rhyming, Jeff Foxworthy–loving ranchers are lefties. The general patterns, though, are consistent and persistent enough to suggest that people's tastes and preferences connect to differences in their political orientations.

Respect My Authority! . . . Or Question It

Social scientists have suspected for the better part of a century that political orientations are tied to personality. Indeed, sometimes the debate has been less about whether personality traits influence ideology than whether ideology *is* a personality trait. Personality can be thought of as the particular patterns of thoughts, feelings, and behaviors that make an individual unique. At first blush, this seems a simple and intuitive concept that we all recognize in ourselves and in others. Johnny is quiet and suspicious and has a temper; Jane is social, curious, and even tempered. The particular bundles of these sorts of traits define us as individuals and are collectively called personality. Few find shocking the claim that our personalities are expressed in the clothes we wear, the cars we buy, the books we read, the careers we pursue, the music we like, the decoration of our homes, the extent to which we are organized, the obedience we display to rules, and maybe even in the enthusiasm we bring to exotic foods and free-form poetry. No less surprising from our perspective is that psychologists and political scientists have produced much research correlating personality traits with broad subjective preferences and also with political attitudes and behavior.

Liberals and conservatives might differ on such eclectic, nonpolitical tastes as literature and salads for the same reason that people with particular personality traits have different tastes and preferences. Maybe "liberal" and "conservative" are just handy terms for describing people who happen to have distinct bundles of traits driving their thoughts, feelings, and actions. The big trick, of course, is figuring out the specific traits that consistently distinguish liberals from conservatives.

If such traits do exist, they should be relatively fixed and stable. Personality is seen as a long-term characteristic, a sort of psychological gyroscope that helps us navigate and interact with our environments, which is another way of saying we are predisposed to favor or reject particular appeals, be they to the latest fashion, the opportunity to get involved in a party game, or the political issue of the day. Context undoubtedly matters. If no one suggests a couple of rounds of Twister, then that party game is not going to reveal people's introverted or extroverted natures. Once that stimulus is present, though, people's responses are predictable. Certain individuals are more likely to be the first to spin the dial and put their right hand on yellow, while others will consistently hang back and watch others entangle their limbs for amusement's sake. Personality tendencies are apparent across a variety of contexts and times.

Of course, we are less interested in how party games sort out introverts from extroverts than whether there are particular personality traits that systematically correlate with political stimuli. Is there a specific set of traits that predispose people toward certain ideological appeals or levels of civic involvement? Are there traits that influence a broader set of tastes and preferences that just happen to shape politics? Social scientists have investigated both possibilities, correlating broad personality traits with political preferences, as well as searching for a set of specific traits that can be thought of as constituting distinct political personality types. The latter approach is perhaps best exemplified by a stream of research that has spent 75 years investigating whether authoritarian orientations are identifiable as personality traits. This research, to put it mildly, has been controversial. It began with reflections on the traits that make for a good Nazi.

Erich Jaensch was a psychologist in pre–World War II Germany who was best known for his work on eidetic imagery. (An eidetic image is an image that is perceived as real but is not.) He began classifying people on their eidetic capabilities and then, ominously, began attaching cultural significance to these capabilities. Somewhere in the process this agenda morphed into providing scholarly cover for some of the more odious racial elements underpinning Nazi ideology. Jaensch's argument that eidetic individuals are more likely than noneidetic individuals to possess certain traits does not seem particularly freighted with political importance. Yet from this basis he started to develop a classification scheme for two personality types that accrued major political ramifications.

The "J" type personality was athletic, practical, and decisive. The "S" type was individualistic, egocentric, and liberal. J-types were likely to be upstanding Nazis; S-types, according to Jaensch, were more likely to be Jews and perhaps Frenchmen. He saw these personality types as biologically (read racially) rooted and not just connected to different views of the world, but maybe even different forms of humanity that would take predictably different sides in any cultural conflict. There is no prize for guessing whom Jaensch viewed as the good guys in such conflict.[13]

After the J-types jackbooted themselves and everyone else into a bloody global conflict and lost, they were viewed less as practical and decisive than as existential threats to humanity. During and immediately following World War II, a number of social scientists investigated the inner workings of J-types. No one really believed in Jaensch's chain of inference—the conclusions were not only morally repugnant but empirically unsupported—but the concept of an authoritarian personality type had been floating around in academic circles for quite a while. Maybe humanity was not divided into J- and S-types that were biologically fated to clash in a global struggle for cultural dominance, but maybe there was in fact something to the notion of certain personality traits being more acceptant of authoritarian social structures. This possibility was a very big deal in the middle of the twentieth century, when authoritarian ideological systems aggressively sought to replicate themselves through persuasion or force.

The big ideological isms that threatened democracy—Fascism on the right and Communism on the left—were clearly aided not just by the acceptance but in many cases the enthusiastic support of large numbers of seemingly ordinary people. This support came despite the indisputable fact that these regimes often fostered scientific ideas that were specious. In addition to Jaensch's extrapolation from eidetic capabilities to justifications of Nazi racial purity, there is the famous example of the Soviet pseudoscientist Trofim Lysenko. Lysenko endeared himself to Stalin and others by rejecting accepted Mendelian genetics in favor of an "anti-bourgeois" agronomy based on a warmed-over belief that acquired traits could be passed along genetically, an assertion that set back science in the USSR by decades and eventually contributed to widespread food shortages.

Moreover, the moral consequences of the policies being justified were difficult to miss: gas chambers and gulags, wars and genocide, manmade famines and a generally cavalier approach to human rights and dignity. What was truly puzzling to social scientists was that these policies were accepted and implemented not only by the ideological true believers but by the average tovarisch and burgher. To be sure, many people just acquiesced in order to protect their families and many others actively resisted at extreme costs to themselves. Yet authoritarian regimes needed average citizens to get with the program. And lots of them did. Why?

In the 1940s and 1950s, psychologists began to hypothesize that people with certain preferences, such as a desire for social order and clear, universally followed rules and regulations, were more likely to provide support to authoritarian regimes. They began to wonder whether these preferences were embedded in deep psychology; in other words, whether they constituted a distinct and identifiable set of traits that could be isolated as a personality type. Thus was born the notion of the authoritarian personality. Investigators working on this topic accepted at least parts of Jaensch's conception of a J-type but viewed such personality types as threats to, rather than foundations of, society. A number of names are associated with the academic work foundational to developing and testing the concept of the authoritarian personality. The most prominent, though, is that of Theodor Adorno, a German academic whose experience with authoritarianism was all too practical. He was a man of broad interests—he is still remembered as a music and cultural critic—who studied philosophy, psychology, and sociology, receiving his PhD in 1924. A rising star in more than one academic field, Adorno fled Hitler's regime after losing his right to teach. Adorno's father was a Jew who had converted to Protestantism, a dangerous genealogy to have in Nazi Germany.[14]

Adorno ended up at the University of California–Berkeley studying, among other things, the sociology and psychology of prejudice. There he began a series of collaborative research projects with Else Frenkel-Brunswik, an Austrian-born psychologist and fellow refugee of the anti-Semitic pogroms of the Hitler regime, and Daniel Levinson and Nevitt Sanford, two psychologists who studied ethnocentrism. These projects resulted in *The Authoritarian Personality*. Theories about authoritarianism had been making the rounds in respectable academic circles for a decade or more before this book was published in 1950,[15] but this was likely the first—and certainly best known—systematic empirical investigation into whether there was such a thing as a personality rooted in politics. As the authors put it, their major hypothesis was "that the political, economic, and social convictions of an individual often form a broad and coherent pattern, as if bound together by a 'mentality' or 'spirit' and that this pattern is an expression of deep-lying trends in his personality."[16] Specifically, they were interested in "potential" Fascists—not overt and committed ideologues but rather those who would be predisposed to support Fascism should it become a mainstream social movement.

They ended up developing the F-scale (the "F" stood for Fascist). This scale focused on nine traits, or "central trends in the person," and for the most part ignored specific policy issues like preferences on hiring quotas. These traits included conventionalism (a rigid adherence to conventional, middle-class values), superstition and stereotypy (a belief in mystical determinants of individual fate coupled with a predisposition to think in rigid categories), and anti-intraception (an opposition to the subjective or the imaginative).[17] A number of the questions they used to try and get at these traits were prescient in that they reflected the sorts of nonpolitical items now known to distinguish between liberals and conservatives. For example, one of the anti-intraception questions was: "Novels or stories that tell about what people think and feel are more interesting than those which contain mainly action, romance and adventure." One of the conventionalism questions asked whether it was more important for a person to be artistic and sensuous or neat and well mannered.[18]

While prescient in its attempt to use nonpolitical questions to tap into the psychology presumed to underlie political beliefs, the F-scale was a bust as a reliable measure of political personality. F-scale scores were predictive of many things, but it was not at all clear what the F-scale was actually measuring, with the potential exception that it was *not* measuring some sort of coherent proto-Fascist personality. The methodological problems of the F-scale were variously attributed to its roots in less empirical, mostly Freudian, psychological concepts; to a somewhat loose approach to picking traits and questions; and, most worryingly, to the motivations of the researchers. A number of modern scholars see the F-scale as saying less about the personalities of those who took the test than about the understandable concerns of its lead authors that a potential Nazi lurked within the average Jane or Joe.[19]

While generally reckoned a failure in terms of generating a reliable measurement scale, *The Authoritarian Personality* is still viewed as a significant and influential piece of scholarship because it sparked broad interest in the concept of political personality types. While quick to find and publicize the flaws in the F-scale, social scientists broadly recognized that Adorno and his colleagues were onto something by viewing politics as an extension of personality. Improvements on the F-scale began to appear. In *The Psychology of Politics*, published just a few years after *The Authoritarian Personality* (1954), Hans Eysenck argued that personality was projected onto social attitudes. In this book, Eyesenck suggested that ideology was a product of two core underlying dimensions. One of these dimensions amounted to a basic left-right take on political and social issues. The other was "tendermindedness" or "toughmindedness." The idea was that ideology depended not just on issue preferences but also on underlying personality. Authoritarians, be they on the left or right, were more likely to be toughminded. Eyesenck put both Communists and Fascists in this category since both groups were willing to pursue their political beliefs with little regard for the preferences and interests of others.

Despite Eysenck's balanced treatment, as this research stream developed it somehow lost interest in one side of the political spectrum, focusing instead almost exclusively on the political views associated with right-leaning politics. For example, in the 1960s Glenn Wilson and colleagues in England, New Zealand, and Australia took the basic concept of conservatism as reflecting a dimension of personality characterized by resistance to change and adherence to tradition. The result was the C-scale ("C" for conservatism) also broadly known as the Wilson-Patterson index—versions of which we use in our own research. They measured conservatism with questions probing attitudes on everything from school uniforms to the death penalty and found it to correlate not just with the political orientations you would expect but also with tastes and preferences more broadly.

A more recent extension of the authoritarian personality research program is right-wing authoritarianism (RWA). It was developed in the 1970s and 1980s by Canadian psychologist Robert Altemeyer, who wondered whether there was a set of people "so generally submissive to established authority that it is scientifically useful to speak of 'authoritarian people.'"[20] Altemeyer's answer was a decisive yes. He spent decades refining what amounted to an RWA personality test. Through various iterations, this test included questions that had clear political implications, but he also experimented with questions that dealt with child rearing, music and films, and personal hygiene. RWA had much stronger psychometric properties than its F-scale predecessor and Altemeyer consistently reported that people scoring high on RWA tests were more likely to support controls on personal freedom and harsh forms of punishment, be hostile to perceived out-groups (e.g., homosexuals and feminists), and be more likely to support government persecution of these groups.[21]

Though academics have spent more than a half-century trying to measure a personality trait tied to some form of preference for authoritarianism, the overall results are mixed.[22] Critics have consistently raised two red flags. One is mostly technical and deals with the statistical and psychometric soundness of all the scales cited above. Are these scales really measuring a set of relatively fixed and stable traits that constitute an identifiable personality type, or are they just proxies for general issue attitudes that do not always hang together statistically?[23]

The second major criticism relates to possible ideological bias. The suspicion is that hiding underneath all the math and psychological theory is a bunch of left-leaning academics bent on identifying a type of personality that is an existential threat to democracy and human decency. And, wouldn't you know it, that personality type tends to consistently correlate with right-leaning political preferences. Though an interest in left-wing authoritarianism surfaces occasionally (Eysenck's work, for example), the major goal has been to identify a personality type that is associated with right-leaning political orientations. Some see this as evidence that conservatism is being held up as a pathology by the left-leaning denizens of academic psychology. One of the most dogged critics of this line of research is psychologist J. J. Ray, an iconoclast who published numerous papers criticizing the technical aspects of this work. Ray, along with other conservatives, believes that "Leftist political tenets . . . form part of the culture of modern-day psychology. When psychologists study ideology, therefore, a tendency to draw conclusions that accord with Leftist beliefs is rather to be expected."[24]

Ideological arguments aside, most of the scales developed to measure politically relevant personality types follow the basic approach of Adorno in that they rely on questions lacking an overt connection to politics. It is possible to go some way toward figuring out how liberal or conservative someone is by knowing the sort of novels they like to read and whether or not they like pajama parties. Even the items in these scales that do have obvious political connections (e.g., adherence to religious tradition and deference to social convention) are distinguishable from the specific issues-of-the-day approach to ideology favored by political scientists. Yet these scales reliably predict specific issue attitudes at different times and in different societies. They tend broadly to correlate with party identification and self-placement on the left-right continuum in different societies at different times. There may not be such a thing as an authoritarian personality, but there is a deep psychology underlying politics. The traits correlating with political attitudes tend to be those that involve attraction to the new, the novel, and the abstract or those involving a sense of duty, order, and loyalty.

Liberoverts

Essentially the same conclusion has been drawn more directly from research that is focused on personality as a general concept, as opposed to a specific political personality type. Though agreement is not universal, modern psychology has posited the widely accepted notion that human personality is composed of five broad underlying traits: openness to experience, conscientiousness, extroversion, agreeableness, and neuroticism. Unlike the personality research stimulated by Adorno, the Big Five model was not motivated by a desire to identify traits associated with politics, but rather as a means for categorizing the broader concept of personality. A number of psychologists working in independent teams managed over the course of several decades to identify clusters of adjectives that tended to go together in describing particular traits, slowly triangulating on more or less the same five-category taxonomy of descriptions. This work was codified into a standardized personality test by scholars such as Lewis Goldberg and the team of Paul Costa and Robert McCrae. This test and its measurement of the five core personality traits along with the resulting Big Five model have been validated across numerous samples and in different societies.[25]

Though the Big Five Personality battery includes nothing overtly political, several items either resemble the nonpolitical probes included in authoritarian scales or tap into the sorts of tastes and preferences discussed earlier. Further, two of the Big Five dimensions consistently correlate with political orientations: openness and conscientiousness. Openness means openness to experience and information and refers to people who are curious, creative, and arty, those who enjoy and seek out novel experiences and are more likely to adopt unconventional beliefs. Conscientiousness means a tendency to be dependable, dutiful, and self-disciplined. On standard Big Five personality tests, openness is assessed by asking people to rate themselves on things like their interest in abstract ideas and whether they have vivid imaginations. Conscientiousness is assessed with ratings on things like paying attention to details and getting chores done.

Numerous studies have linked these personality dimensions to differences in the mix of tastes and preferences that seem to reliably separate liberals and conservatives. People who score high on openness, for example, tend to like envelope-pushing music and abstract art. People who score high on conscientiousness are more likely to be organized, faithful, and loyal. One review of this large research literature finds these sorts of differences consistently cropping up across nearly 70 years of studies on personality research. The punch line, of course, is that this same literature also reports a consistent relationship between these dimensions of personality and political temperament. Those open to new experiences are not just hanging Jackson Pollock prints in disorganized bedrooms while listening to techno-pop reinterpretations of Bach by experimental jazz bands. They are also more likely to identify themselves as liberals. High conscientiousness types are not just hanging up patriotic posters in neat and tidy offices while listening to their favorite elevator music. They are also more likely to identify as conservatives. These relationships hold up across time, across societies, and in studies using a wide variety of conceptual and methodological approaches.[26]

The connection between conservatism and conscientiousness is consistent with a substantial body of research indicating that people with a great desire for what is known as "cognitive closure" are more likely to be politically conservative. For two decades, scholars have employed a collection of survey items such as "I think that having clear rules and order at work is essential to success," "I do not like situations that are uncertain," "I like to have friends who are unpredictable," and "Even after I've made up my mind about something, I am always eager to consider a different opinion."[27] Higher values indicate a stronger preference for closure (a dislike of uncertain situations and less interest in entertaining alternative viewpoints, for example). Comparing political orientation to scores on these items reveals that in a variety of countries, individuals who are fond of closure also tend to self-identify as conservative, vote for traditional parties, and favor conservative positions on both social and economic issues.[28] Perhaps not surprisingly, a fondness for closure also correlates with religious fundamentalism.[29]

None of this should be taken to imply that conscientiousness and comfort with authority structures and clear answers are viewed as character flaws. To the contrary—given the chance to describe themselves in survey self-reports, many people are eager to overstate their degree of conscientiousness. Millions of people display these traits while still being able to recognize a morally compromised authority structure when they see one. Bear in mind that a personality characterized by openness does not predetermine a liberal political temperament any more than being conscientious means you are predestined to be a conservative (think probabilistically). The key point is that the openness and conscientiousness survey items do not include "political" questions, yet they persistently correlate with political orientations, suggesting that something deep in human psychology predisposes people to a broad variety of likes and dislikes that guide thoughts, feelings, and actions. Some of this shows up in taste for music, for art, for clarity, for salad greens, for politics, and perhaps for morality.

On What Foundation Is Your Morality Built?

Moral foundations theory is a project of social psychologists trying to figure out why moral norms vary with culture, yet still seem to reflect certain human universals. All cultures have seemingly idiosyncratic notions of what is right and wrong, yet clearly there also is evidence of universal ethics. The morality of polygamy, infanticide, racism, and sexism changes across cultures and across time. The morality of incest and murder does not. What gives? Moral foundations theory, developed primarily by psychologists Jonathan Haidt and Jesse Graham, argues that moral universals are rooted in "intuitive ethics." This is the notion that all humans come equipped with a set of innate psychological mechanisms that automatically trigger emotionally based moral responses to the situations we encounter in our physical, psychological, and social environments. Peel off the academic language and the core idea is one we all recognize. Most everyone experiences gut responses to ethical dilemmas (Should I go back and tell the cashier he gave me too much change? Should I rat out the person who is stealing from the supply cabinet?). In these sorts of situations an inner voice often tells us, "Go make things right with the cashier because otherwise he's going to be in trouble when his register doesn't balance."

Haidt and Graham argue that these sorts of ethical intuitions are based in five distinct universal systems—termed moral foundations—that account for the vast majority of moral decision-making across cultures. They identified these foundations by asking people to reflect on the concerns relevant to them when they determined whether something was right or wrong. Two of these foundations deal with the unjust treatment of individuals, specifically whether someone was harmed or treated differently from others (respectively labeled the harm and the fairness foundations). The other three foundations focus less on the individual than on the group or community and deal with loyalty and betrayal, respect for authority, and the desire to avoid that which is vile and disgusting (labeled loyalty, authority, and purity, respectively).[30] From differential emphasis on particular sets of foundations, different ethical systems can be constructed individually and socially.

Though moral foundations theory starts from a different place than the trait-based personality research—the search for a universal ethics as opposed to the search for the qualities that form individual character—it ends up in pretty much the same place, at least in terms of politics. This is because the moral foundations you use to decide what is wrong and what is right are fairly accurate predictors of your political beliefs. What Haidt, Graham, a number of collaborators, and an expanding set of independent investigators find is not just evidence that these moral foundations are identifiable across cultures, but also that they are indicators of political temperament. The quick summary is that liberals tend to place their emphasis on the foundations relating to the unjust treatment of individuals (harm and fairness) while conservatives are likely to rely more heavily than liberals on concerns for loyalty, authority, and purity. In other words, when it comes to deciding what is the morally correct course of action, liberals are particularly sensitive to the way in which an individual is being treated, while conservatives are more likely to factor in group considerations.

A liberal likely sees a moral wrong when an individual is being, say, socially ostracized. A conservative is more likely to take into account communal considerations in formulating a moral judgment. Is that guy being ostracized because he is not one of us? Because he was disloyal? Because he broke the rules or thumbed his nose at the accepted way of doing things? Because he did something that everyone else finds disgusting? If the answer to these sorts of questions is yes, maybe he had it coming. One of the important implications of moral foundations theory is that liberals and conservatives disagree not because they have rationally analyzed their way to different issue positions but rather because they have different reflexive responses to what is going on in their social, psychological, and physical environments. These responses are emotionally rooted cues to what is right and what is wrong. Our term for these emotionally rooted, reflexive responses is predispositions.[31]

Haidt and Graham describe moral foundations as the "taste receptors of the moral sense,"[32] and it is worth noting how these moral taste buds line up with the personality research. Emphasis on harm and fairness is positively associated with liberal orientations, positively associated with the personality trait of openness, and negatively associated with right-wing authoritarianism (RWA).[33] This makes sense; people high on openness are more likely to be sensitive to anything that constrains individual expression and freedom. People with high conscientiousness and RWA are more likely to be sensitive to anything that violates group-oriented rules and regulations. Like personality, moral foundations seem to be capturing something universal—a set of dispositions that guide reactions to situations in our physical, social, and psychological environments. Among other things, those dispositions clearly influence our politics.

Moral foundations theory also helps to explain why conservatives and liberals differ on tastes and preferences that seemingly have nothing to do with politics. For example, Haidt and Graham asked people what sort of dogs they wanted. Liberals wanted dogs that were gentle and related to their owners as equals. Conservatives wanted dogs that were loyal and obedient. These pet preferences mapped directly onto differences in underlying moral foundations, with liberals emphasizing traits associated with just treatment and conservatives emphasizing traits reflecting loyalty and authority.[34]

Like moral foundations theory, values theory is another framework that backs up the general conclusions drawn from the work on personality traits. Values theory was developed to investigate broad and universal aspects of human psychology. Values in this case refer to enduring goals and are indicative of the aspects of the world that motivate people's beliefs and actions. Psychologists and cultural anthropologists have long been intrigued by the possibility that humans share a core, universal value system, and probably the best-known theory encompassing this idea was developed by Hebrew University psychologist Shalom Schwartz.

The foundational assumption of values theory is that individuals in all societies must be responsive to three things: biologically based needs (like the need to eat); social needs (like the need to communicate with others); and group needs (the need to secure the welfare and survival of the group). Regardless of time or place, people must figure out what has to be done to meet these needs and, as social animals, need to be able to collectively coordinate on this all-important to-do list.[35] These needs form the basis of value systems and Schwartz distilled from these 3 basic needs 10 broad values. Each of these values is distinguished by a central motivational goal that is linked to one of the three core needs. For example, one of the core values is hedonism, which reflects a goal of individual pleasure and gratification (think eating, drinking, and sex) and obviously promotes survival. The value of conformity, in contrast, reflects a goal of social maintenance, a motivation to keep your hands off other people's plates as well as their partners and not just as a means to avoid getting slapped but to serve the larger goal of maintaining social order. Researchers have found these 10 core values, and have validated their motivational content in more than 70 distinct cultural groups.[36]

Clearly, these values do not always work in harmony: Hedonism is individualistic and conformity is communal. Individuals have widely varying belief systems, with some emphasizing group over individual welfare and others the opposite. Interestingly, Schwartz and his collaborators not only show that

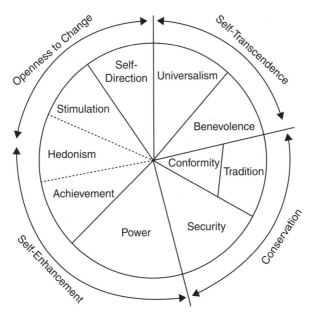

Figure 4.2   Schwartz Typology of Basic Human Values

the 10 core values are identifiable in a wide variety of societies and cultures, they demonstrate that they almost always clump together on two dimensions. One of these dimensions reflects individual (achievement and power) versus collective welfare (universalism and benevolence). The other dimension distinguishes group loyalty (security, conformity, and tradition) versus do your own thing (pleasure, novelty, excitement, and independent thought). As shown in Figure 4.2, if these two dimensions are placed on top of each other, the result is four big "slices," which have been labeled "self-enhancement," "self-transcendence," "openness to change," and "conservation," and the position of the values in this two-dimensional space reflects their theoretical and statistical relationship with each other.

You don't have to stare at the pie too long to notice that the 10 basic values approximate an ideological continuum. On one half are motivational goals that relate at an individual level to novelty, creativity, and freedom to do your own thing, and, at a group level, to taking care not just of your own group but of everyone everywhere. On the other half is a set of values that represent an individual level motivational goals for working hard and getting ahead and, on a group level, to tradition, security, and conformity. These "pie slices," in other words, are not simply expected to correlate with differences in political temperament; they purport to explain why those differences exist. According to values theory, political orientations spring from the different motivational biases that underlie the taxonomy of values. A number of studies find empirical support for this hypothesis. One study, for example, looked at left-right orientations in 20 countries and found universalism and benevolence (the self-transcendence pie slice) consistently associated with a left-leaning orientation and conformity and tradition (the conservation pie slice) with a right-leaning orientation.[37] These value dimensions also correlate with Big Five personality traits in the way you would expect, especially on the key factors of openness and conscientiousness.[38]

The distance from salad greens to moral foundations and values is considerable. In fact, if Haidt and Schwartz are to be believed, we have entered the realm of ethics and the sources of beliefs and actions. Yet even as we have descended into the depths of human psychology, the scenery has not changed much in that the psychological survey items consistently correlate with political orientations as referenced by the left-right ideological divide. The left is characterized more by a desire for the new and novel, a commitment to individual expression, and a tolerance of difference; the right by a desire for order and security, a commitment to tradition, and group loyalty. These differences correlate with disagreements over the best dish to have for dinner. They also correlate with disagreements over the best individuals and political parties to run the government.

Politics Has an Odor

Evidence exists that a range of seemingly nonpolitical tastes and preferences correlate with political temperament. Liberals and conservatives consistently differ in the way they answer a variety of survey questions and these differences persist over time and across cultures. The diversity of the questions eliciting politically relevant responses is astonishing; they range from occupational preferences to leisure pursuits to sensitivity to disgust, as well as personality traits, moral foundations, personal values, culinary choices and preferences for music, art, cars, humor, poetry, fiction, neatness, and all the other concepts discussed above. Yet all this says little about the underlying mechanisms at work. Why do any of these variations map onto political orientations?

For some political scientists the answer is that political views come first and then somehow spill out into broader psychology and tastes.[39] We are skeptical. For the sake of argument, though, let's accept the basic premise. Assume that liberals explicitly or implicitly encourage their children to crave exotic foods and artistic novelty. Assume that conservatives see other conservatives being neat and orderly, enjoying poems that rhyme, and reading novels that bring clear closure to the plot, and follow suit. These assumptions still beg the question of why political liberals/conservatives display these particular packages of life preferences in the first place. Was it just a chance occurrence that somehow perpetuated itself (path dependence)? If we could rewind the clock and start again, would we be just as likely to end up with liberals rather than conservatives preferring realistic art, conscientiously picking up their rooms, and looking forward to meat and potatoes for dinner? This possibility seems highly unlikely. For one thing, this scenario is wholly inconsistent with the finding that the same political orientations align themselves with the same personal preferences, personality traits, moral foundations, and personal values regardless of the culture and time. The patterns are simply too consistent to attribute them to some prehistoric protoliberal developing a taste for arugula and serving it at protoliberal gatherings, thereby somehow setting in motion a string of events that would culminate in a global left-wing salad green cult.

If it is not political beliefs driving overall life orientations, an obvious alternate causal argument is that these broad life orientations are driving our political beliefs; indeed, this is the basic assumption of much of the research discussed above. Personality traits/moral foundations/personal values come first and they provide the source of partisan affiliations, ideological orientations, and issue positions. Accepting this argument, though, still does not provide a full explanation for why some people are more open to new experiences, more favorable toward individualism in moral judgments, and more interested in values such as self-expression. A third causal explanation, and the one we tend to prefer, is that bedrock political orientations just naturally mesh with a broader set of orientations, tastes, and preferences because they are all part of the same biologically rooted inner self.

To get a rough notion of the relevance of biology to tastes, we return to Zev Witsotsky's beer-loving flies from the beginning of this chapter. The University of California–Riverside team identified the biological basis for glycerol fondness in fruit flies in a gene (Gr64e) that shaped this element of gustatory response. Flies lacking a particular form of that gene preferred sugar water to beer. What this clearly demonstrates is a biologically embedded predisposition—specifically a taste preference shaped by gustatory receptors. Could the same thing happen in humans? Well, yes.

The genetic basis of human flavor preference was discovered in 1931 by Arthur Fox, a chemist working for Dupont. As is so often the case with discoveries, it was in part an accident. Fox was pouring a powdered form of the compound PTC (phenylthiocarbamide) when some of the stuff became airborne. A colleague remarked that the PTC floating around the room tasted bitter. Fox was puzzled because he couldn't taste a thing. He and his colleague started experimenting to see who could and could not taste PTC. At the annual meeting of the American Association for the Advancement of Science in 1931, they had audience members take a PTC taste test and found that some people were extremely sensitive to the taste of PTC, some could taste it but weakly, and some could not taste it at all. Later, the researchers were able to establish that the ability to taste PTC is traceable to a couple of dozen genes that shape individual variation in whether and how strongly PTC is tasted.

PTC taste strips are now a staple of grade school science classes designed to demonstrate variation in taste as well as the genetic basis of that variation. As plenty of science class alums can attest, those who can taste PTC generally experience it as a bitter flavor. This is interesting because those extremely sensitive to PTC also tend not to like vegetables that contain bitter secondary compounds such as those found in vegetables like broccoli and, yes, arugula.[40] You can probably guess what is coming. Does PTC taste sensitivity systematically correlate with political ideology? One of our graduate students—Jayme Neiman—empirically tested that proposition on a group of undergraduates and found that conservatives are significantly more likely than liberals to detect PTC.[41]

Interesting explanatory possibilities emerge. Food preferences are one of the nonpolitical traits that consistently (if modestly) differentiate liberals and conservatives. Food preferences are certainly shaped by culture and family but they are also incontrovertibly rooted in biology. Taste buds, after all, are nothing more than chemical receptors and people differ markedly in the density and nature of these chemical receptors. It does not get more biological than that. The relevance of genes to taste further diminishes the attractiveness of the argument that the reason little conservatives hate broccoli is solely because they are taught to do so by their conservative, broccoli-hating parents, and raises an interesting question. Are genes relevant to tasting PTC merely linked (close by on the same chromosome) to genes relevant to politics, or is the relationship more meaningful? Again, we should not get too carried away here. Taste preference for a single compound, regardless of how genetically shaped, is likely to explain only a small amount of our culinary, let alone political, preferences. Besides, taste is only one sense. Variation in other senses is unlikely to connect to politics, right?

Evolutionarily speaking, olfaction is the most ancient and chemically direct of our sensory systems; its signals register more or less directly in the emotional centers of the brain. Olfaction encodes a good deal of information concerning what is good in our environments (the smell of a good meal) and what is bad (the smell of rot or decay). Olfaction is also known to carry a good deal of social information. For example, how attractive you are to the opposite sex depends partially on the olfactory compounds you are emitting, which is the basis of the $10 billion personal fragrance industry. Some of the social information wrapped up in a sense of smell relates not just to figuring out if someone is sexy but to their position in a social hierarchy, and this is getting close to politics.

Might variations in sense of smell be related to preferences for authoritarian or egalitarian social structures? In one of our own studies, we found evidence in support of such a possibility. An odor of great potential interest to social life is that emanating from androstenone, a nonandrogenic steroid closely related to testosterone (testosterone itself emits no odor). Some people smell androsterone readily and report it to be overpowering; others cannot detect it at all. Further, among those who do smell it, there is variation in the odor's appeal: Some say it smells like sweat or even urine, but others say it has a pleasant odor, like incense, sandalwood, or vanilla. A long-established and well-replicated connection exists between various forms of a particular gene (OR7D4) and the ability to smell androstenone. We found that variations across people in the ability to detect the odor of androstenone predicts political attitudes such that those more sensitive to the odor of androstenone are also more comfortable with clear social hierarchies.[42] The relationship might be spurious and definitely needs to be replicated before it is accepted with certainty, but the possibility is intriguing. We know that the world smells differently to some people than to others, and variations in the ability to smell androstenone might be related to political beliefs.

Conclusion: Taste Buds, Olfactory Bulbs, and Politics

The key question of this book is: What makes conservatives and liberals different? Our answer is that they experience and process different worlds. In this chapter, we argued that biology (which, as we shall see soon enough, is much more than just genetics) predisposes people to certain preferences and tastes because the individual differences discussed in this chapter extend to biology. Each person experiences the world differently because the biological machinery responsible for that experience—the sensory, perceptual, and processing systems—differs from one person to the next. We taste and smell the same things differently. We cognitively and subjectively interpret the same paintings or stories or jokes differently. We have different personalities, moral foundations, and personal values—and we have different politics.

Notes

[1]   Wisotsky et al., "Evolutionary Differences in Food Preference Rely on Gr64e, a Receptor for Glycerol."

[2]   Zeleny, "Obama's Down on the Farm."

[3]   Hunch.com, "How Food Preferences Vary by Political Ideology"; and Hunch, "You Vote What You Eat: How Liberals and Conservatives Eat Differently."

[4]   These surveys can all be found at neuropolitics.org.

[5]   Since 1992, *Family Circle* magazine has asked the spouses of the major party presidential nominees to submit cookie recipes, and readers vote for their favorite. The winner of this contest has predicted the presidential winner in four of five elections. The exception was in 2008, when Cindy McCain's oatmeal-butterscotch cookies got the nod over Michelle Obama's shortbread. This win was controversial, though, as McCain was accused of plagiarizing the winning recipe.

[6]   Wilson et al., "Conservatism and Art Preferences"; and Dollinger, "Creativity and Conservatism."

[7]   Gillies and Campbell, "Conservatism and Poetry Preferences."

[8]   Wilson, "Ideology and Humor Preferences"; and Wilson and Patterson, "Conservatism as a Predictor of Humor Preferences."

9   Carney et al., "The Secret Lives of Liberals and Conservatives: Personality Profiles, Interaction Styles, and the Things They Leave Behind."

10   For example, see Rothman et al., "Politics and Professional Advancement among College Faculty."

11   Tierney, "Your Car: Politics on Wheels."

12   Leder, "What Makes a Stock Republican?"

13   Diamond, "The Co-Ordination of Erich Jaensch."

14   Probably the best biography of Adorno is Stefan Muller-Doohm's *Adorno: A Biography*.

15   It was not just Nazi apologists like Jaensch doing this sort of theorizing. Those who saw fascism as a global threat to freedom of democracy were also openly speculating about what made people turn to and support authoritarian regimes. See, for example, Erich Fromm's *Escape from Freedom*, first published in 1941.

16   Adorno et al. *The Authoritarian Personality*, 1.

17   Ibid., 222.

18   Ibid., 228.

19   See, for example, Martin, "*The Authoritarian Personality*, 50 Years Later: What Questions Are There for Political Psychology?"

20   Altemeyer, *Right-Wing Authoritarianism*, 7.

21   A related concept and survey battery goes by the name of social dominance orientation (SDO). It was developed by psychologists Jim Sidanius and Felicia Pratto, who argued that it represents a personality trait reflecting the degree of an individual's preference for inequality among social groups. People with high SDO scores tend to be driven, meritocratic, and tough minded, and tended to display low levels of empathy and tolerance and to be highly supportive of policies that enforce group-based inequality. Across multiple samples, SDO was found to correlate with nationalism, patriotism, conservatism, and racism. Pratto et al., "Social Dominance Orientation: A Personality Variable Predicting Social and Political Attitudes."

22   Probably the best-known contemporary research on the political relevance of authoritarianism is being done by Vanderbilt political scientist Marc Hetherington and various colleagues. In a 2009 book, Hetherington and coauthor Jonathan Weiler argue that authoritarianism provides a particularly compelling explanation of the current polarization of American politics. See Hetherington and Weiler, *Authoritarianism and Polarization in American Politics*.

23   One of the most damning criticisms of the early work on the authoritarian personality (up through Wilson's C-scale) was penned by Altemeyer, who devoted roughly a hundred pages of his 1981 book to a thorough dismantling of this research.

24   Ray, "The Scientific Study of Ideology Is Too Often More Ideological Than Scientific."

25   Goldberg, "The Structure of Phenotypic Personality Traits"; and Costa and McCrae, *NEO PI-R: Professional Manual*. For a good history of the development of the personality research that led to the Big Five model, see Digman, "Personality Structure: Emergence of the Five-Factor Model."

26   A good summary of these findings is found in Carney et al., 2008; see Table 1, 816. Some studies not included in their summary that reinforce the same point using samples from different nations at different times and using a variety of methods include Neiman, "Political Ideology, Personality, and the Correlations with Tastes and Preferences for Music, Art, Literature, and Food"; Feist and Brady, "Openness to Experience, Non-Conformity, and the Preference for Abstract Art"; Rawlings et al., "Personality and Aesthetic Preference in Spain and England: Two Studies Relating Sensation Seeking and Openness to Experience to Liking

for Paintings and Music"; Furnham and Walker, "The Influence of Personality Traits, Previous Experience of Art, and Demographic Variables on Artistic Preference"; Furnham and Avison, "Personality and Preferences for Surreal Art"; Mondak et al., "Personality and Civic Engagement"; Gerber et al., "Personality and Political Attitudes"; and Mondak, *Personality and the Foundations of Political Behavior.*

27   Kruglanski et al., "Motivated Resistance and Openness to Persuasion in the Presence or Absence of Prior Information."

28   See, for example, Golec, "Need for Cognitive Closure and Political Conservatism: Studies on the Nature of the Relationship"; Kossowska and van Hiel, "The Relationship between Need for Closure and Conservative Beliefs in Western and Eastern Europe"; Chirumbolo et al., "Need for Cognitive Closure and Politics: Voting, Political Attitudes and Attributional Style"; Federico et al., "The Relationship between the Need for Closure and Support for Military Action against Iraq: Moderating Effects of National Attachment"; and Jost and Kruglanski, "Effects of Epistemic Motivation on Conservatism, Intolerance, and Other System Justifying Attitudes."

29   Linesch, "Right-Wing Religion: Christian Conservatism as a Political Movement"; and Streyffeler and McNally, "Fundamentalists and Liberals: Personality Characteristics of Protestant Christians."

30   Haidt and Graham, "When Morality Opposes Justice: Conservatives Have Moral Intuitions That Liberals May Not Recognize." The list of foundations in moral foundations theory has undergone quite a bit of refinement and development, and the dimensions listed in the text may travel under different labels in different publications. There are also candidates for a sixth or even seventh moral foundation—liberty/oppression and waste. A summary of the literature on Moral Foundations Theory and its extensions and updates is available at MoralFoundations.org.

31   Though Haidt's argument is less based in biology than ours, the general notion of political attitudes and behaviors being based in reflexive emotionally laden responses is clearly compatible with our conception of predispositions as biologically instantiated "defaults" that influence political temperament, often outside our conscious awareness. See Haidt, *The Righteous Mind.*

32   Haidt et al., "Above and below Left-Right: Ideological Narratives and Moral Foundations."

33   Ibid.

34   Haidt, *The Righteous Mind,* 187–188.

35   Schwartz and Bilsky, "Toward a Universal Psychological Structure of Human Values."

36   Schwartz, "Universals in the Content and Structure of Values: Theoretical Advances and Empirical Tests in 20 Countries"; and Schwarz and Boehnke, "Evaluating the Structure of Human Values with Confirmatory Factor Analysis."

37   Olver, "Personality Traits and Personal Values: A Conceptual and Empirical Integration."

38   Jang et al., "Heritability of the Big Five Personality Dimensions and Their Facets: A Twin Study."

39   Verhulst et al., "Correlation Not Causation: The Relationship between Personality Traits and Political Ideologies."

40   Krebs, "The Gourmet Ape: Evolution and Human Food Preferences."

41   Neiman, "Phenylthiocarbamide Detection and Political Ideology." Paper presented at the annual meeting of the ISPP 35th Annual Scientific Meeting, Chicago, IL, 2012.

42   Smith et al., "Political Orientations May Vary with Detection of the Odor of Androstenone."

CHAPTER 5

Do You See What I See?

Tell me to what you pay attention and I will tell you who you are.

Jose Ortega y Gasset

If by a "Liberal" they mean someone who looks ahead and not behind . . . then I'm proud to say I'm a Liberal.

John F. Kennedy

In 1918, Hermann Rorschach was a young psychiatrist working at an asylum in Herisau, Switzerland. The job had its enlightening moments. One of his research interests involved analyzing the leader of a small religious sect who believed his penis was sacred and should be adored by followers. Rorschach's interest in phallocentric prophets—and aren't they all—was soon derailed by a larger fascination with inkblots. This interest probably traced to Rorschach's childhood, which was spent in Zurich, where a popular children's pastime was "klecksography." This was an arty activity that involved putting a dab of ink on a page and folding it in half. Voila, the smudged ink takes the form of a discernible object. Butterfly wings, for example. Rorschach, though, was not interested in using inkblots to make pictures that people could see. He was interested in what people could see in pictures made from inkblots.[1]

Rorschach's big contribution to psychology is, of course, the test that bears his name and continues to be given to millions around the world. It is what's known as a "projective" test. The basic idea is that you ask someone to identify an ambiguous stimulus such as an inkblot and his or her response will reveal something about that person's underlying psychology. The meaning and validity of the test as an analytical and diagnostic tool has been a matter of some debate since Rorschach first published his arguments and findings in 1921. Nonetheless there is no doubt that people really do see different things in those inkblots and that many in the field of psychology and psychiatry have treated those differences as indicative of an individual's personality traits, cognitive processing patterns, and perceptual orientation to the world.

Rorschach tests are often associated with Freudian psychoanalysis, but don't worry. We have no intention of speculating on the potty training of conservatives or the mommy issues of liberals. We are interested, though, in the core empirical finding of millions of Rorschach inkblot tests: Give people the same visual stimulus and they will respond differently. They see different things and pay attention to different things, and the pieces of information sieved out by these contrary perceptual screens are processed into different conclusions and beliefs about their environment. In a series of creative experiments, psychologists have elaborated on this basic fact. What these experiments fairly convincingly demonstrate is that people have different patterns of attention, information processing, and decision-making. We wondered whether those differences might somehow systematically relate to political temperament. It turns out they do. Differences in political temperament are tied to differences in a variety of perception and processing patterns prompted by stimuli. In other words, liberals and conservatives may, quite literally, see the world differently.

The Eyes Have It

Imagine you agree to participate in a social science experiment. The next thing you know you are seated in front of a standard-issue computer screen that has a standard-issue computer keyboard in front of it. The instructions are straightforward: "Please press the space bar on your keyboard as soon as a large black dot appears on the screen. Ignore anything else on the screen because it will be completely irrelevant to the location of the dot." Before any black dot makes its appearance, a round, cartoonish face materializes in the center of the screen. The eyes of this face are looking to its right, your left. Next, a large black dot appears

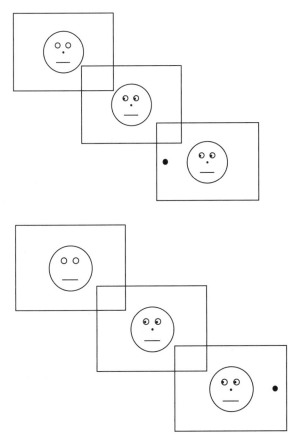

Figure 5.1 Schematic of Gaze Cuing Test

in the approximate portion of the screen where the cartoon face's eyes are looking. You give the space bar a bash as per your instructions. The sequence repeats itself, except this time the dot appears to your right, opposite from the direction of the face's eyegaze. This goes on for a while: Sometimes the eyes look right, sometimes left; sometimes they look toward where the dot pops up and sometimes not. Figure 5.1 gives the general idea of the process. This all seems pretty silly and maybe you would rather be back in Chapter 3 with the guy asking you to divvy up a free 20 bucks. There is a reasonable chance, though, that you have just demonstrated an interesting pattern: The eye gaze of the face influences where you look even though you know your task is to respond to the black dot, and even though you have been told that the face and eyes provide no clues as to where the dot is going to appear.

Numerous studies have shown that people are, on average, significantly quicker to hit the space bar when the dot appears in the direction of the cartoon face's gaze.[2] Obviously, people are "cued" by the gaze and their attention drifts in that direction. This works when the cue is being delivered by real faces or by cartoon faces.[3] People, at least on average, seem to be sensitive to social cues such as the direction of eyegaze.[4]

That phrase "on average" just appeared again and by now we hope that whenever it does it makes you wonder about individual differences. Who wasn't average, and why? What was the biggest cuing effect registered by an individual? What was the smallest? Do particular types of people tend to be more influenced by cuing effects? True to social science form, research on gaze cuing has not been that interested in individual variation.[5] This is a shame because the extent to which individuals are susceptible to gaze cues is indicative of important information regarding their sensitivity to social cues and more generally their thinking patterns. On those occasions when individual variation has been addressed, it is typically to compare clinical and nonclinical groups. For example, autistic individuals are largely unaffected by gaze cues, a result perfectly compatible with data showing autistics typically avoid eye contact and have "theory of mind" deficiencies.[6] There has been much less research on what accounts for the substantial variation within groups of research subjects who are in the "normal" range; that is, people who are not diagnosed with one malady or another. Yet even among those people who have never been checked against the *Diagnostic and Statistical Manual* (DSM), some are much more susceptible to cuing effects than others. We were intrigued by this largely unexplored variation. Specifically, we wanted to know whether liberals are systematically different from conservatives in their susceptibility to gaze cuing effects.

Research we conducted in collaboration with psychologist and vision specialist Mike Dodd suggests the answer is yes. We lured a large group of undergraduates into Mike's lab with promises of extra credit and gave them the gaze-cuing task represented by Figure 5.1. Averaging all students together, we found the basic cuing effect; overall, they were slower to hit the space bar when the cartoon face's gaze was in the opposite direction of the dot. Our main interest, however, was in the variation around that average. Were liberals systematically more susceptible to the gaze cuing effect than conservatives? We suspected this would be the case and we turned out to be correct. Our data showed that the half of the sample holding the most liberal positions on a range of important political issues were much slower to hit the space bar when the gaze was misdirected. In fact, they were really slow (by the way, we're talking about measuring the speed of basic mental processes, so "really slow" in this context means 22 milliseconds slower).

On the other hand, a misdirected gaze had basically no impact on the more conservative half of the sample; they were slowed by only 3 milliseconds, a mere one-seventh of the observed impact for liberals. Since the effect for conservatives was statistically insignificant, we can conclude that, for all intents and purposes, gaze cuing did not influence them. That was a bit of an eye opener— so much so that we wondered if we had an unusual sample. So we replicated the study on a nonstudent sample and found a gap between liberals and conservatives that was still statistically significant and in the same direction. Liberals seem to be much more influenced by gaze cues than conservatives—a finding that is buried if only the average size of the gaze cuing effects is reported.[7]

Our findings suggest that liberals are more influenced by social cues—even when told to explicitly ignore those cues. Conservatives seem to be more willing or able to ignore cues and follow the rules that govern the situation ("the lab proctor says I should ignore the gaze, so I will ignore the gaze"). There is clearly a difference here and the temptation is to figure out what it implies in a normative sense. Is the tendency to be influenced by social cues a character flaw or a point of pride? Are those who ignore extraneous information to be commended or lamented? Are liberals better than conservatives? Are conservatives better than liberals? These kinds of questions are irrelevant to us—and certainly inappropriate for us to answer—because they have no objective answer. From one perspective, it could be argued that, at least in some situations, ignoring social information is beneficial because those around us often provide behavioral cues that are useless or even dangerous. Countless parents hammer this point home with "just because Tommy/Tammy wants to play in a busy street does not mean it is a good idea" speeches. On the other hand, it can also be argued that being in touch with those around us is a central and deeply appealing aspect of the human condition. Joint attention of this sort is an important component of interaction,[8] a good, natural, binding, and necessary step toward understanding and empathy. Both attentional styles, in other words, have pros and cons.

Declaring one as preferable to the other is an impossible task because people have different standards regarding desirable behavioral patterns. In fact, we suspected that liberals and conservatives not only behave differently in the gaze cuing task but think differently about the preferred or appropriate behavioral response to this sort of social cue. Before we did any behavioral tests, we asked our subjects the following somewhat odd question: "Some people's eyes just naturally tend to look where others are looking, but other people tend not to be influenced by where others are looking. All else equal, which do you think is the better way for people to behave, look where others are looking or not be influenced by where others are looking?" People split pretty evenly, with 55 percent asserting that it is better to look where others are looking and 45 percent saying that it is better not to be influenced by where others are looking. Correlating these responses to a measure of conservatism, we found that conservatives were more likely to believe it is "better" for people not to be influenced by where others are looking, while liberals were just the opposite. As usual, the correlation coefficient is not large ($r = 0.14$), indicating that many conservatives believe it is better to look where others are looking and many liberals believe it is better to ignore where others are looking. Still, this relationship was statistically significant, so we can place some degree of confidence in the pattern. Liberals and conservatives not only respond differently to gaze cues, they think people *should* respond differently.

Given that the standard used to judge the desirability of a particular behavior varies across people and across political ideologies, global judgments about the desirability of a particular way of responding to gaze cues are just not possible. It is better simply to conclude that liberals differ from conservatives in important ways and that, not surprisingly, each behaves in the fashion they deem more appropriate. Liberals and conservatives not only play the game of life differently, they apply different rules for evaluating play.

Fitting Round Pigs into Square Holes

Is a giraffe a zoo animal or a farm animal? What about a pig? Though the answers to these questions are pretty obvious, other animals are harder to classify. Bison and llamas, for example, are seen on farms and in zoos. Geese are rarely found either place. Categorizing a goose or a buffalo as either a farm animal or a zoo animal is not obvious or easy—reasonable people will classify them differently. Similarly, is a Barbie doll a toy or not? What about a lawnmower? Categorizing either of these items as a toy or not a toy seems easy—but this is not the case for other objects such as trampolines, water balloons, and high-powered pellet guns. Certainly children play with trampolines, water balloons, and pellet guns, so they could be categorized as toys. Yet trampolines can also be thought of as gymnastic equipment, water balloons as tools of mischief, and pellet guns as weapons (all three of us had mothers who emphasized that our spanking new, pump-action, .177-caliber slayer of sparrows was "not a toy!").

In an extremely creative doctoral dissertation, Everett Young concocted a large number of categorization tasks. He asked research participants to sit at a computer and, by a click of the mouse, place sequentially presented target objects into one of two large squares labeled, for example, zoo animals and farm animals. Young also gave participants the additional option of placing the target object between the two categorization squares if they felt that the target "belonged simultaneously to both categories" or was "between the two labeled categories." Young's interests were not really focused on whether the general consensus these days is that a buffalo is a zoo or farm animal, but rather in people's tendency to place objects in provided categories as opposed to a middle ground of some sort. Young calls those who dutifully place targets in the provided categories "hard categorizers" and those who opt for the middle ground "soft categorizers," and it turns out that people who tend to be hard categorizers for animals also tend to be hard categorizers for toys, home appliances, and healthy versus unhealthy foods; the same cross-topic consistency applies to soft categorizers.

Young found that this individual tendency toward hard or soft categorization—the tendency to take what we see and divide it into clearly labeled and separated mental boxes or, alternatively, to opt regularly for the mental equivalent of a catchall kitchen drawer—turns out to be a pretty good predictor of political temperament. Liberals are more likely to be soft categorizers and conservatives hard categorizers. This pattern holds up whether liberalism is measured by the participants' self-classification as a liberal or a conservative, by their preferences on economic issues, or by their preferences on social issues. Again, the correlation coefficients are not overly strong—in the 0.10 to 0.20 range—but they are consistent and statistically significant. Young is understandably proud of these results, noting that "the abstractness of the [categorization] tasks" provides better evidence than survey items of "a real cognitive-process precursor to ideological thought."[9] Variation in the categorization tasks indicates that differences in how people process information and in what they do mentally with what they see are intimately intertwined with political orientations.

Like the gaze cuing results as well as the results on conscientiousness, openness to new experience, authoritarianism, and preference for cognitive closure, the categorization results suggest that conservatives are more likely than liberals to lock on to a task and complete it in a fashion that is both definitive and consistent with instructions. Liberals are more likely to be distracted, to equivocate, and to be flexible even to the point of not performing the task exactly as the authorities intended. Are liberals and conservatives different in their cognitive orientations? Definitely. Is it preferable to be cognitively oriented like liberals or like conservatives? Your answer to this question is probably hopelessly influenced by whether you are a liberal or a conservative.

Our Thoughts Are Our Own—or Are They?

What people are paying attention to would seem to be an internal matter. As teachers, we know it is sometimes possible to tell whether a student is daydreaming or listening, but for the most part, thoughts are simply unavailable to the people who are not thinking them. This is a key issue for psychologists because they get paid to poke around in that inner world to see what is going on. In order to do so, they have designed some devilishly clever ways of figuring out where people are directing their attention without relying on the people themselves to provide that information verbally. Three common tools psychologists employ for this sort of thing are the Stroop, Dot-Probe, and Flanker tasks.

The Stroop Task is familiar to many people. It consists of a set of words presented one at a time to a research subject who is then asked to report the color of the font in which each word appears. Reading a target word, however, naturally directs attention to the word itself rather than to the color of the font, a situation that is especially telling when the word is a color. For example, when the word "blue" appears in red font, research participants take longer to answer "red" because they are reading the word "blue." What's more interesting is that, with some slight modifications, the Stroop Task can be employed to figure out the types of words to which different people naturally pay attention. The way this works is that target words are presented in random order, with some appearing in red font and some in blue. Each target word appears at least twice, once in each color. As before, the subject's task is to ignore the target word and, as quickly as possible, report the color of the font in which the target word is written. The longer it takes to report font color, the more that word is grabbing the attention of the participant. In this way, the Stroop Task can be employed to figure out if people pay attention not just to single words but to categories of words. Some people may be diverted by action words and others by object words; some by words dealing with human relationship and others by words dealing with contests of skill; some by words associated with sex and others by words associated with food.

A group of Italian researchers led by Luciana Carraro wanted to know why some people tended to pay more attention to negative than to positive words. To find out, they presented a Stroop Task to a set of undergraduates at the University of Padua. Some of the target words were positive, as in the top line of words below, and some were negative, as in the bottom line. Carraro's team found that some research participants were much slower in reporting the font color of negative as opposed to positive words and wondered what explained this variation.

pleasure	paradise	wonderful	freedom	security	happiness	joy
sickness	horror	pain	contempt	suffering	evil	dirty

Weeks earlier, they had asked the students to report the extent to which they agreed (from 1 = not at all to 7 = very much) with six politically charged topics in Italy: reduction of immigration, abortion, medically assisted procreation, same-sex marriage, use of arms for personal defense, and adoption by same-sex couples. After rescaling responses so that higher numbers always represented conservative positions and then adding the six items together, this measure of political preferences was compared to response-time differences in the subsequently administered negative-positive Stroop Task. The correlation was positive, sizable ($r = 0.38$), and statistically significant, suggesting that more conservative individuals tend to have "stronger automatic vigilance toward negative as compared to positive stimuli."[10]

Conservatives and liberals were about equally quick and accurate in reporting the font color of the entire group of target words, but when broken down
by categories of words, liberals took basically the same amount of time regardless of whether the target was a negative or a positive word. Conservatives,
on the other hand, were significantly slower when the words were negative.
Participants were also asked to evaluate the positive and negative words from
1 (extremely positive) to 7 (extremely negative), and liberal and conservative students were no different in their evaluations of these words. Thus, the
researchers concluded that political ideology is not related to explicit evaluations of stimuli but it is related to "automatic allocation of attentional resources."

This difference in attention to negative stimuli does not just hold for words
but also for images, as demonstrated by the same Italian researchers using a
Dot-Probe Task. Here, two different images appear simultaneously on the computer screen, one on the left side and one on the right. After a half second or
so, a gray dot appears on one of the pictures. The participants are instructed
to report as quickly as possible whether the dot appeared on the picture on the
right or the picture on the left. By measuring how long it takes participants to
note the location of the dot, researchers can figure out which image the respondent was paying attention to before the dot's appearance. The basic logic here
is simple. Suppose in one case the two images appearing were a snake and a
flower. If an individual is much quicker to identify the location of the gray dot
when it appears on the image of the snake than when it appears on the image
of the flower, the respondent's attention was probably already tilted toward the
snake when the dot appeared.

The researchers selected eight positive and eight negative images from
the International Affective Picture System (IAPS), which is a large data set of
images that have been previously rated (on scales ranging from favorable to
unfavorable and from arousing to not arousing) by participants who had nothing to do with our projects.[11] Some pictures, such as those of fruit bowls, happy
people, and cuddly animals, are consistently rated much more favorably than
others, such as dangerous animals, accidents/disasters, and bodily wastes. Pairs
of images—one negative, one positive—were presented to research participants
(another group of Italian undergrads), with the grey dot popping up with equal
frequency on the nasty and nice pictures. By looking at how much faster (or
slower) people were to find the grey dot on a positive versus negative image, the

researchers were able to measure how much attention was being devoted to one sort of image versus the other.

The basic finding was that liberals were a bit quicker on the draw when the dot popped up on a positive image, while conservatives were a bit faster when the dot popped up on a negative image. In fact, the correlation of speed differential with political orientation was sizable ($r = 0.36$). As with the words in the Stroop Task, when participants were later shown the 16 images and asked to rate them, liberals and conservatives did not differ much. In other words, the difference was in their attention to different types of images rather than their evaluations of those images.

The Italian studies, like all others, have some drawbacks. The research participants were all college undergraduates and the particular political issues used to assess political temperament primarily deal with social issues such as homosexuality, immigration, and gun control. College students tend to lean liberal on these sorts of issues, even if they are more conservative on others. These limitations, though, are more than counterbalanced by the fact that the participants were not from the United States. Most of the research on the deeper bases of differences in political predispositions has been conducted in the United States, leading some (especially those who do not believe politics has cross-cultural commonalities) to suspect that the results in other contexts would not be parallel. The Italian findings clearly show that for at least one non-U.S. population, the attention of political conservatives tends to be directed at negative as opposed to positive words and images. So the obvious question is whether the pattern observed in Italy is cross cultural. For example, does the tendency of conservatives to devote more attention to negative stimuli also apply to U.S. participants?

With a group of psychologists, we conducted a preliminary study to address this question, and the answer seems to be yes.[12] Our results were derived from yet another exercise employed in psychology—this one called the Flanker Task. Here, participants are told to identify some feature of a target image appearing in the center of the screen—for example, its color or degree of pleasantness. The trick is that when the target image eventually appears it is "flanked" by two other images. Research consistently shows that when the two flanking images are incongruent with the target image (perhaps a different color or differentially pleasing), participants typically take longer to complete the assignment they were given. Individuals' attentiveness to the target image can be measured by recording how much they are slowed down by incongruent flanking images. The less participants are slowed down, the more attention they were devoting to the target image.

The particular variant of the Flanker Task we employed utilized faces expressing either anger or happiness. Participants saw four categories of images: (1) an angry target with angry flanking images, (2) an angry target with happy flanking images, (3) a happy target with happy flanking images, (4) and a happy target with angry flanking images. We told our subjects to ignore the flankers and, as fast as possible, to press one of the two keys indicating a happy or an angry target as soon as an image appeared. Like the Italian researchers, we used college undergrads as our research subjects and also used issue positions to create an index of conservatism.

We found that our undergrads showed the same sorts of general "flanker effects" that have been demonstrated in dozens of previous studies. For example, as a whole they were quicker to respond to an angry target than a happy target, regardless of whether the flanking images are happy or angry. This attentional bias towards angry faces has long been noted and is probably a product of the fact that, from an evolutionary point of view, people sensitive to angry faces might be more likely to survive than those without much of a sense of what angry expressions looked like.[13] It is all well and good to note those in our vicinity who are happy, but from an evolutionary point of view it is absolutely essential to note those who are angry. Happy people might pat you on the back; angry people might stab you in it. We also found that our undergrads demonstrated the most standard and widely replicated flanker effect—they were quicker to identify a target when the flanker traits were the same as the target trait.

Those are all average effects, though, and the key issue for us was whether variance in the flanker effect correlated with political orientation—especially whether conservatives display less of a flanker effect when the target is angry. This is indeed what we found. When the target is an angry face, conservative participants focus so much on it that whether the flanking images are the same or different is largely irrelevant. The opposite is *not* the case when the target is a happy face. Liberals and conservatives both display the traditional flanker effect in this situation. In other words, they are slower to identify the target as happy when the flankers are angry. It is only when the targets are angry that the flanker effect more or less disappears, and then only for conservatives. Conservatives, in short, seem to be more likely to lock their attention on the negative (angry) stimulus and the negative stimulus only.

So what we have here are three different research paradigms, three different samples, three different types of visual stimuli (words, images, and facial expressions) conducted in two different countries that all lead to the same conclusions. Compared to those holding liberal political beliefs, conservatives tend to direct more attention to negative stimuli. Again, we are not ready to proclaim the existence of an Iron Law of Conservative Negative Attention—many replications need to be done and there are surely nuances yet to be discovered and investigated. For starters, it would be nice if we could directly measure what people are looking at rather than infer it based on response times. And actually, we can.

What Are You Looking At?

The technology to follow the precise track of the eyes has existed for some time. In the nineteenth century, scholars attempted to learn more about the reading process simply by watching people's eyes as they read. Not satisfied with the level of precision of this technique, in the very early twentieth century, Edmund Huey devised a contraption that was not much more than a lightweight aluminum pointer stuck to a contact lens; the pointer moved when the eye moved. Crude as it was, Huey's device helped demonstrate that as people read their eyes do not fixate on every word in a sentence.[14] A couple of decades later, Guy Buswell developed the first nonintrusive eyetracker that, in essence, filmed light reflecting on the eye.[15]

Eyetracking technology has improved dramatically in the last 100 years, and these days the majority of eyetrackers utilize infrared light that is reflected by the eye and then picked up by an optical sensor. This technology constitutes a fairly easy and noninvasive way to gather accurate data on eye movement, and it is a big bucks business. Advertisers, web designers, marketers, and product development specialists can manipulate you much more effectively thanks to their knowledge of common eyetracking patterns. The same technology also has more practical and even altruistic applications, helping disabled individuals send emails, text messages, and browse the Internet using only their eyes. Training simulators of all sorts rely on eyetracking technology to improve the skills of novice drivers, assist the military, and train commercial pilots. Fatigue-detection equipment often is based on eyetracking technology and is making the roads and skies safer. Geriatric research using eyetrackers found that walking problems of the elderly are often vision problems. A variety of training techniques for athletes and for adolescents have benefited from eyetracking technology. Finally, academics have also gotten into the act since what people are eyeballing turns out to say a lot about underlying attention and related cognitive processes. As psychologist Alfred Yarbus once said, "[E]ye movement reflects the human thought process."[16]

That seemed like something of interest and it just so happened that our aforementioned colleague Mike Dodd runs an eyetracking lab. With his help we decided to test whether people with distinct political orientations have distinct eyetracking patterns when presented with the same images. To do this, we showed a group of 76 undergraduates collages of four images taken from the IAPS repository, with one image presented in each quadrant of a computer screen. Examples of positive images included a skier having fun, a beautiful sunset, a happy child, an arrangement of fruit, a cute rabbit, and a beachball. Examples of images that rated negatively included a person with a mouthful of worms, an open wound, a shark with teeth bared, a wrecked car, and a house on fire.[17]

Each of the collages had a mixture of positive and negative images. In this particular study we compared six collages that had three negative and one positive image with six collages that had one negative and three positive images (an example of a collage with three negative and one positive image is presented in of Figure 5.2). We showed each participant the exact same set of collages with each collage on screen for 8 seconds. Participants were told to look anywhere they wanted as long as it was on the screen (this is called "freeview"). They did not need to hit a space bar, look for dots, or check for colors. We were interested in the length of time participants looked at particular images in the collage. This is known in the business as "dwell time." Our central expectation was that conservatives would spend more time checking out the negative—potentially threatening—stimuli.

We measured political temperament by asking participants about their party identification, issue stances, and general attitudes on leadership, treatment of out-groups, adherence to traditional norms of behavior, and other aspects of social order and maintenance (in other words, the bedrock issues in our measure of ideology). These diverse measures were combined to provide

Figure 5.2 Sample Collage of Three Negative and One Positive Image

an overall indication of each participant's location on a liberal to conservative index. To make the results easily presentable, we divided the participants at the halfway point of this index and presented averages for the 38 above this point (conservatives) and the 38 below it (liberals).

Overall, our results revealed that participants spend more time looking at negative rather than positive images. This is consistent with previous research and with evolutionary selection pressures that weed out dimwits too focused on the berries on the bush to notice there's a bear in there, too. Our real topic of interest, however, is whether, compared to liberals, political conservatives tend to spend more time looking at the negative images. Indeed they do. Figure 5.3 shows the "dwell time" differential for these two groups of participants. The upshot is that liberals spent a little more time (400 milliseconds or less than half a second) looking at negative than positive images. Conservatives spent a lot more time (a bit over 1.5 seconds) looking at negative compared to positive images.

In the context of an 8-second freeview exercise, this size of "difference in the differences" is huge—indeed, one vision specialist referred to it as an "eternity."

Total dwell time is only one of several useful measures that can be derived from eyetracking data. Another is time until first fixation. Before each collage was presented, our participants were instructed to look at a focus point (an "X") in the center of an otherwise blank screen. When the collage of four images appears participants typically scan the screen before "fixating," or stopping, on one image another. Thus, we can precisely measure how quickly a participant fixates on a negative as compared to a positive image after the collage pops onto the screen. The results of first fixation time for the collages are reported in panel 2 of Figure 5.3.

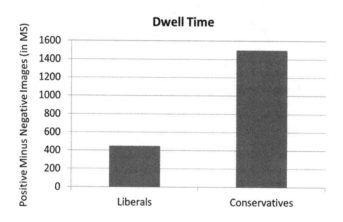

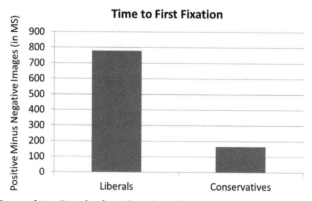

Figure 5.3   Eyetracking Results for Liberals and Conservatives

Remember, the quicker fixation time, the greater attentional bias toward negative images. Note how quickly conservatives fixate on the negative image, an average of about 165 milliseconds. Liberals take more than half a second longer to fixate on the negative (again, a pretty big difference given the context and the speed of the underlying mental processes). We also found a difference between liberals and conservatives in the length of time before they fixate on a positive image, in that liberals were quicker to focus on positive stimuli in their environment. The findings, though, were considerably weaker than what we found for conservatives focusing on negative images, leading to the conclusion that liberals were quicker to focus on the good stuff, but not by much.[18]

Worried that undergraduates might be atypical, we repeated the study on a random sample of over 100 adults. The results were essentially the same or just a tad weaker. The bottom line is that our eyetracking study triangulates straight onto the same basic inference taken from time response tasks such as the Stroop, Dot-Probe, and Flanker: People who devote more attention to negative stimuli tend to report having conservative political orientations.

Perception Is Reality—but Is It Real?

When they are forced to look at the same thing, liberals and conservatives pay attention to different aspects of that thing, but nothing thus far suggests that they are perceiving the images differently. Even though conservatives focus more on the negative images, liberals and conservatives appear to perceive negative images as similarly negative and the positive images as similarly positive. At least, this was the conclusion of the Italian researchers. Remember, they found that when participants were shown the same positive and negative IAPS images (the ones used in the Dot-Probe Task) and asked to report whether those images were extremely positive (1), extremely negative (7), or something in between (2 to 6), liberals evaluated them in a fashion roughly equivalent to conservatives. The same lack of any difference extended to the positive and negative words used in the Stroop Task. Does this mean liberals and conservatives really see the same world but only pay attention to different things? We're not so sure.

Certainly they perceive different political worlds. Everybody knows that liberals and conservatives have different political preferences—surveys make this rather obvious point all the time. Conservatives prefer this policy, while liberals prefer the opposite. But note that what is being asked here is what policies are *preferred*. We decided to do something different. In one of our surveys, we asked a random sample of respondents to forget about their preferences and provide their *perceptions* of current public policies. Specifically, we asked them to locate on a scale of 1 to 10 their view of the status quo on six policy dimensions: tolerance of new rather than support for traditional lifestyles; of leaders who are decisive and firm rather than cautious and open to dissent; of policies that do everything possible to protect against external threats; of policies that strictly punish rather than display compassion for rule-breakers; of policies that benefit the rich even if they are undeserving rather than benefit the poor even if they are not making an effort; and of a government that is only minimally involved in society rather than a government that is involved in nearly all facets of life.

On each of these six dimensions we found that conservatives and liberals see the operative policies and practices in the United States very differently. Not surprisingly, conservatives are more likely than liberals to perceive the country as having policies that tolerate new lifestyles, do little to protect against outside threats, mollycoddle criminals, and benefit the poor even if they are not making the effort. In particular, the difference with regard to the perceived treatment of the rich and the poor in the current United States was huge. In other words, it is not just that liberals and conservatives prefer different policies; they see different policies currently in place. Liberals see current policies benefiting the undeserving rich. Conservatives look at those same policies and situations and see the undeserving poor with their snouts in the public trough. Other surveys confirm these basic differences.[19] This is important—it suggests that when liberals and conservatives look through the same window they see different worlds. We think it is these differing *perceptions* that are a major source of political conflict, not just the difference in *preferences*. Republican presidential nominee Mitt Romney's comments on the "47 percent who pay no income taxes" and will never vote Republican are indicative of a perception widely held by conservatives and, as soon as the comments came to light, liberals let him know that their perceptions of the "47 percent" were quite different.

These perceptual differences are not limited to politics. We asked a sample of U.S. adults to rate a series of IAPS images on a scale of 1 (favorable) to 9 (unfavorable). Some of these images were decidedly positive, including a cuddly baby, a beautiful image of Niagara Falls, and individuals enjoying outdoor sports. Others were threatening—pictures of knife and gun attacks on vulnerable people. Others were disgusting and included a very used toilet, open wounds, vomit, and an emaciated body (our research subjects get their participant fees the old-fashioned way—they earn them). In contrast to what the Italian researchers found, our data suggest that conservatives perceive generic negative images more negatively than do liberals. Considering the disgusting images as a group, the correlation between unfavorable evaluations and conservative views on issues is positive and significant. The same pattern holds for threatening images, though the relationship is slightly weaker. Conservatives appear to rate negative images a bit more unfavorably than do liberals.[20]

A surprising twist appears, however, when the collection of positive images is analyzed. People holding conservative views on political issues tend to rate the positive images even more favorably than liberals, so it is not the case that conservatives perceive everything more negatively.[21] A more accurate interpretation is that, compared to liberals, conservatives attach greater emotional punch to whatever stimulus they are presented with—rating the positive images more favorably and the negative more unfavorably.

A related finding comes from psychologist Jacob Vigil's research on facial expression processing.[22] He reasoned that people who support the political party that typically supports "dominant responses to domestic and international conflicts" (in the United States that would be Republicans) are likely to be those who are quick to interpret various environmental stimuli as threatening. One of the most evolutionarily ancient forms of social communication is facial expression, so Vigil hypothesized that an ambiguous stimulus in the form of a neutral facial expression would be more often perceived by Republicans, compared to Democrats, as threatening or dominant. Threatening facial expressions include anger, fear, or disgust, while nonthreatening expressions include joy, sadness, or surprise. Dominant expressions are generally reckoned to be joy, anger, or disgust, while submissive expressions include sadness, fear, or surprise. As can be seen, according to this generally accepted categorization scheme, threatening expressions are usually but not always dominant, and nonthreating expressions are usually but not always submissive. The exceptions to the pattern are that joy is dominant but nonthreatening, and fear is threatening but submissive.

To test his hypothesis, Vigil had one male and one female actor portray five ambiguous facial expressions. Research participants (over 800 undergraduates at the University of Florida) were asked to identify the face as expressing sadness, joy, disgust, surprise, fear, or anger. He found that compared to students sympathizing with the Democratic Party, Republican sympathizers were more likely to interpret the faces as signaling both threatening and dominant emotions.[23] This relationship seemed to persist even after statistically controlling for standard demographic variables as well as psychosocial variables such as trust, life satisfaction, and aggression.

The results on differing perceptions may be slightly less clear than those for differing patterns of attention, but evidence exists that conservatives perceive disgusting images more unfavorably than do liberals, that they perceive threatening images slightly more unfavorably than liberals, and that they perceive positive images more favorably than liberals. Vigil's study suggests that Republican undergraduates may be more inclined than their Democratic peers to see an emotionally ambiguous face as threatening, or more likely dominant. Thus, it would seem that conservatives and liberals do not perceive stimuli in exactly the same way. If that is really the case, the big issue is whether these sorts of differences extend beyond attention and perception and into the way people acquire and use information.

You're Full of Beans

Psychologist Russell Fazio and his colleagues are interested in how people acquire and use information. To help investigate and understand those processes they developed a game called BeanFest. (See figure 5.4, page 137.) It works like this. Participants sit at the computer and are presented with a picture of a bean. Then another bean and then another. The beans come in different shapes, from circular to oblong, and with different marking patterns, from one dot to many dots. As a bean is presented, the player decides whether to accept or reject the bean. Each type of bean is worth a point value of either +10 or –10 and the participant does not know the value of the bean when he or she first encounters it. For example, an oblong bean with two tightly spaced dots in the middle may be worth –10 and a circular bean with three dots in a triangle may be worth +10. If accepted, the bean's point value is revealed and

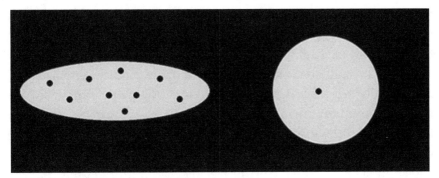

Figure 5.4   BeanFest Bean Examples

the participant's point total is adjusted according to that value. If the bean is rejected, the participant's point total remains the same and the point value of the unselected bean type remains unknown—a lack of information that has consequences for future decisions if the bean is seen again. Participants begin with 50 points and the object of BeanFest is to get to 100. To give participants a bit more incentive, they were promised a monetary payoff if they ran their total to 100.

It did not take long for the researchers to note that people varied widely in the way they played BeanFest. Some threw caution to the wind and accepted beans with abandon. This meant they gained and lost a lot of points but also collected substantial amounts of information about the value of the various beans. Others were much more wary, accepting just a few beans at first and then only accepting subsequent beans that matched the few types known to be good. In other words, some participants were quite a bit more exploratory than others. What accounted for this variation from individual to individual? The researchers first suspected psychological and personality characteristics. Perhaps people who are more open to new experiences and who like to cogitate are more exploratory. Perhaps those more conscientious and eager for closure will be avoidant.

These expectations were usually borne out, but they are not too surprising— this is pretty traditional psychological stuff. Then, quite by accident, Fazio and his coauthor Natalie Shook learned that a group of their BeanFest participants had been involved in a separate study that included a measure of political ideology. On a lark, the researchers checked for a correlation between approach behavior (in the context of the game, this means accepting new beans) and political liberalism. They found a fairly strong correlation, ($r = 0.25$), enough to convince them to do an original study of politics and BeanFest. This study confirmed that self-identified politically conservative participants approached (accepted) fewer beans than liberal participants. In fact, the correlation was a bit stronger ($r = 0.30$) than the serendipitous finding. Identifying who does and does not turn over virtual beans to discover point totals might seem like just another mildly interesting I-can't-believe-they-get-paid-to-do-this social science study, made notable only for its possibilities as a source of flatulence jokes. The implications, though, are potentially far reaching; they suggest liberals and conservatives not only have different preferences and see things differently, but process what they see differently. Shook and Fazio see the results as indicative of differences in data acquisition strategies and learning styles. Confronted with the opportunity to acquire new but potentially bothersome information some people decline, and these individuals tend to be politically conservative. Others will forge ahead and damn the consequences. These individuals tend to be politically liberal.

The differences identified by BeanFest do not stop with decisions on whether or not to accept a new bean; they extend to differences in the way liberals and conservatives learn. Shook and Fazio had the foresight to include an important follow-up to the BeanFest experience itself. After the participants made their decisions on whether to accept or decline each of the many beans and after the overall point total was tallied, they were asked to take a final exam. Each type of bean was presented in turn and the participant was asked to report whether the bean was a good one (+10) or a bad one (−10). Surprisingly, given the fact that they were acquiring significantly less information than liberals (because they were declining more opportunities to learn about different bean types), conservatives did not perform any worse than liberals on this exam. There were, however, significant differences in learning and these differences were evident only if positive information was analyzed separately from negative information. Liberals were just a bit better at remembering which beans were bad than they were at remembering which beans were good; however, there was no such approximate balance for conservatives—they knew a bad bean when they saw one. Actually, they knew a bad bean even when they didn't see one. Conservatives were way better than liberals at correctly identifying bad beans, but they were also more likely to miscategorize good beans as bad. This is potentially a big deal. What it suggests is that while liberals and conservatives, on average, might get the same scores on the exam, they had learned very different things.

This finding helps to bring together several threads important to our arguments in this book. Conservatives' relative discomfort with the new and unfamiliar shows up not only in self-reports about themselves but in behavioral patterns like a reluctance to acquire new but potentially risky information. Such reluctance has pros and cons; it protects conservatives from negative situations but also means that invalid negative attitudes cannot be disproven ("I haven't seen that bean before, so it must be bad"). This tendency could help to foster the impression that "the world is a relatively harsh place."[24] The perception created by this sort of cognitive process encourages greater restraint in exploring new situations, creating a classic feedback loop. For their part, liberals could be soliciting so much information without any particular focus that they do not process and retain it as well as they could. Shook and Fazio believe that the differential sampling behavior of liberals and conservatives is responsible for this learning asymmetry. New bean? What the hell, say the liberals, let's give it whirl. New bean? Whoa there fella, say the conservatives, better give that one a miss. We're not just talking about beans, of course. Shook and Fazio are suggesting these patterns are stable and generalizable, not limited to a lab-based computer game. If that is true, variation in people's willingness to explore new objects and situations may be at the core of the differing worldviews of liberals and conservatives.

We want to emphasize that liberals neither won the game more often nor received higher scores on the final exam. The key point is, once again, that liberals and conservatives are different. Conservatives acquire the information they believe necessary to draw adequate conclusions, then call it a day. Liberals go on acquiring new information even if they don't like what they find and even if they might not be able to fully absorb all the information they keep collecting. These different strategies lead to different types of learning that shape or reinforce differing worldviews. This line of thinking is consistent with additional evidence compiled by the Italians from a few pages ago.[25] They found that if you provide people with negative information about unacquainted individuals in a small group (for example, "James borrowed a CD from his friend and knowingly did not return it"), conservatives are more likely than liberals to form negative impressions. Again, we see that negative information is learned and weighted differently by liberals and conservatives, which has implications for differences in individuals' default perceptions of unfamiliar groups.

As Shook and Fazio conclude, the evidence points to "fundamental differences in how individuals with varying ideologies approach their social world, acquire information, and form attitudes."[26] Those who are still having trouble grasping how broad psychological and cognitive tendencies can profoundly affect the formation of relatively narrow political attitudes need look no further than variations in "exploring" behavior. People who seek out new information are simply much more likely to arrive at different political conclusions than those who are comfortable avoiding the risk and uncertainty accompanying new information.

The differing orientations to new information are likely to manifest themselves in differing attitudes toward science and religion, with liberals eager for more data even if those data are alarming (think global warming) and conservatives more likely to be content with knowledge that they believe has already been revealed to them. Seen from this vantage point, it is not surprising that attacks on science are more likely to come from the political right.[27] The one-study-shows-this-but-another-shows-that nature of the scientific process is probably more bothersome to the conservative than to the liberal mindset. From the conservative perspective, referring to a set of findings and claims as "just a theory" could hardly be more damning; it bespeaks an absence of certainty that is troubling, especially if someone is proposing big and expensive changes on what is taken to be little more than debatable conjecture. To liberals, theories, even if dissent is present and i's are left undotted and t's uncrossed, are much more valuable—the weight of current scientific evidence is likely good enough for them and future modifications to knowledge (look, a new bean!) are more likely to be taken in stride. "The great thing about science is not that it is right but that it can be wrong." Whether you nod your head in agreement with this aphorism or wonder how the potential to be wrong could possibly be advantageous says a good deal about your attitudes toward the acquisition of knowledge . . . and toward the bedrock dilemmas of politics.

Conclusion: To Take Threats Seriously or Not, That Is the Question

Pay attention to stimuli that signal potential threats; follow instructions unless the source is obviously bogus; avoid unfamiliar objects and experiences; keep it simple, basic, clear, and decisive. According to results of the cognitive tasks summarized in this chapter, this tends to be the modus operandi of those who endorse conservative political stances. Liberals seem to follow a different set of guidelines in navigating, perceiving, and understanding their environments: Seek out new information even if you might not like it; follow instructions only when there is no other choice; embrace complexity; engage in new experiences even if they entail some risk.

The stream of research summarized in this chapter suggests that variation in the collections of strategies guiding each individual's choices consistently correlate with traits, tastes, preferences, and political orientations. People who are uncomfortable in situations that are unpredictable and unfamiliar are less likely to travel to new places and to attend parties where there will be numerous strangers. Personality traits obviously affect the kinds of daily situations toward which people gravitate, including the places they choose to live. People desirous of new experiences, unexpected sights, novel foods (arugula, yum!), noise at all hours, and diverse people are likely to end up living in different neighborhoods than those who want homogeneity, quiet, and security. People seem to accept the connection of these behavioral patterns to personality traits—and they should accept their connection to political traits too.

In modern life, political systems have potentially powerful effects on the living patterns of people. Governmental decisions influence perceived internal and external security, acceptable lifestyle options, and population diversity. Conservatives are more likely than liberals to support public policies that will mitigate the dangers of bad things in part because their cognitive patterns lead them to be more cognizant than liberals of bad things. In truth, relative to politics, personal choices of where to live, where to work, and what leisure activities to pursue probably are much more relevant to people meshing their cognitive orientations and psychological preferences with the environments in which they place themselves. Politics, however, is largely a series of collective rather than personal decisions and it just may be that people feel so strongly about politics because it places particular aspects of the environment outside of their control. Many people (but not all—another instance in which individual variation is fascinating and meaningful) find this lack of total control bothersome and for them politics is serious business. People want to be able to shape their surroundings to match their psychological and cognitive styles and politics diminishes their ability to do so. In fact, in nondemocratic systems, politics obliterates this ability; in democratic systems, it places people only partly at the mercy of their fellow citizens.

To return to a couple of our favorite cautions, applying general adjectives to those individuals with certain collections of political beliefs only makes sense to those people who are able to think probabilistically. Fail to understand what a correlation of 0.15 or 0.20 (or even of 0.30 or 0.35) really means and you fail to grasp the message we are offering. So just in case you have forgotten one of the big points of Chapter 1, we'll repeat: There are numerous exceptions to the patterns we identify. None of the correlation coefficients in the studies summarized in this chapter approach 1.0, or 0.75, or even 0.50. Remember, exceptions are plentiful for all social science results, as well as most in the natural sciences, and the results we report are no different.

Modest correlations, though, do not detract from the remarkable consistency found across samples, designs, and tasks. This combination of replication and consistency makes us pretty confident the general conclusions presented rest on solid ground. On the whole, compared to political liberals, conservatives are more desirous of security, predictability, and authority, but that does not mean this is necessarily true of Mitt Romney, Jon Huntsman, Ayn Rand, David Cameron, Aunt Belinda, or you. On the whole, compared to political conservatives, liberals are more comfortable with novelty, nuance, and complexity but that does not mean this is necessarily true of Barack Obama, Harry Reid, Nancy Pelosi, François Hollande, Uncle Louie, or you. In other words, finding exceptions to the general patterns is not going to do much to convince us that those patterns are any less real. For that to happen, there have to be more exceptions than there are people for whom the pattern applies. We can't rule this out, but it will take a good deal of additional systematic research with countervailing results to do that.

Finally, besides the caution about rules and exceptions, we especially want to hammer home the point that any judgments about which cognitive patterns are "best" is subjective. Nothing in this chapter points to an objectively preferable set of life strategies. Indeed, each collection of strategies we have identified has pros and cons. The true answer about which approach to life is better depends on the extent to which new situations and experiences are dangerous, and as we have just seen, the answer to this question is itself subjective. So our advice is to be careful in drawing inferences from findings like those in this chapter. It may seem as though conservatives are being portrayed as pessimists and liberals as optimists. But it is possible to keep an eye on the negative without being a pessimist. In fact, conservatives consistently are found to be more optimistic than liberals—even when controlling for differences in income and social status.[28] Similarly, despite the fact that liberals score high on indicators of hedonism and sensation seeking, they consistently are more empathetic than conservatives.[29] In other words, do not read more into these results than is there.

In our quest to understand the deeper differences characterizing variation across the political spectrum, we have seen in Chapter 4 that liberals and conservatives report distinct personality and psychological tendencies and have different tastes in all sorts of things from art and sports to personality traits and vocational preferences. In this chapter we have seen that liberals pay attention to the eyegaze cues of other people and conservatives pay attention to all sorts of stimuli that are negative. Conservatives' cognitive patterns reveal a comfort level with clarity and hard categorization while liberals are more likely to value complexity and multiple categories. Conservatives minimize negative results by eschewing exploratory behavior, thereby avoiding surprise and occasional disappointment; liberals take chances and attend to and learn about both positive and negative stimuli.

Now it is time to see if these differences between liberals and conservatives extend even more deeply, not just to the way they answer survey items or to the attention they devote to different categories of events and stimuli, but to differences in their physical beings. Could it be that liberals and conservatives have distinct neuroanatomy and biological responses to stimuli? Research on the physiological differences of liberals and conservatives is not as extensive as research on the psychological and cognitive differences, but in Chapter 6 we gather up that which exists and summarize it.

Notes

[1]   Pichot, "Centenary of the Birth of Hermann Rorschach." See also Rorschach's biography entry on whonamedit.com.
[2]   Friesen and Kingstone, "The Eyes Have It! Reflexive Orienting Is Triggered by Nonpredictive Gaze"; and Driver et al., "New Approaches to the Study of Human Brain Networks Underlying Spatial Attention."
[3]   Friesen et al., "Attentional Effects of Counterpredictive Gaze and Arrow Cues"; and Bayliss and Tipper, "Gaze and Arrow Cueing of Attention Reveals Individual Differences along the Autism Spectrum as a Function of Target Context."
[4]   Interestingly, in the animal world, sensitivity to such cues seems to correlate with domestication rather than mental firepower. Dogs are more likely than some species of monkeys to take pointing and gaze cues (Miklosi et al., "Use of Experimenter-Given Cues in Dogs"). This willingness to go along can have a downside: Chimps are better than human infants at ignoring instructions that are obviously superfluous to the task being attempted (Horner and Whiten, "Causal Knowledge and Imitation/Emulation Switching in Chimpanzees and Children").
[5]   Not surprisingly, cuing effects vary depending upon the length of time between the appearance of the face and the appearance of the dot and are biggest when the delay is approximately half a second, long enough for the participant to notice but not long enough for the information to become old hat (Dodd et al., "The Politics of Attention: Gaze-Cueing Effects Are Moderated by Political Temperament").
[6]   Bayliss and Tipper, "Gaze and Arrow Cueing of Attention Reveals Individual Differences along the Autism Spectrum as a Function of Target Context." Gender differences have also been reported by Bayliss et al. See "Sex Differences in Eye Gaze and Symbolic Cueing of Attention."
[7]   Dodd et al., "The Politics of Attention: Gaze-Cueing Effects Are Moderated by Political Temperament."
[8]   Moore and Dunham, *Joint Attention: Its Origins and Role in Development.*
[9]   Young, "Why We're Liberal; Why We're Conservative," 331.
[10]   Carraro et al., "The Automatic Conservative: Ideology-Based Attentional Asymmetries in the Processing of Valenced Information."
[11]   Bradley and Lang, "The International Affective Picture System (IAPS) in the Study of Emotion and Attention."
[12]   McLean et al., "Applying the Flanker Task to Political Psychology."
[13]   Van Honk et al., "Selective Attention to Unmasked and Masked Threatening Words: Relationships to Trait Anger and Anxiety."
[14]   Huey, *The Psychology and Pedagogy of Reading.*
[15]   Buswell, *How Adults Read.*
[16]   Yarbus, "Eye Movement and Vision," 190.
[17]   We used only those images pre-rated in the top 20 percent in terms of negativity and those in the top 20 percent of positivity. The images in Figure 5.3 are not the actual images used

since, in order to preserve the value of these images for future researchers, those who use IAPS must agree not to publish any of the images.

18 Dodd et al., "The Political Left Rolls with the Good and the Political Right Confronts the Bad: Connecting Physiology and Cognition to Preferences."

19 Mitchell et al., "Side by Side, Worlds Apart: Liberals' and Conservatives' Distinct Perceptions of Political Reality."

20 The correlation between unfavorable evaluations of disgust images and conservative issue positions was $r = 0.12$, $p < 0.05$. The comparable correlation for threatening images was $r = 0.10$, $p < 0.10$.

21 The actual correlation was $r = -0.15$, $p < 0.01$.

22 Vigil, "Political Leanings Vary with Facial Expression Processing and Psychosocial Functioning," 550.

23 Ibid., 552.

24 Shook and Fazio, "Political Ideology, Exploration of Novel Stimuli, and Attitude Formation," 3.

25 Castelli and Carraro, "Ideology Is Related to Basic Cognitive Processes Involved in Attitude Formation."

26 Shook and Fazio, "Political Ideology, Exploration of Novel Stimuli, and Attitude Formation," 2.

27 Mooney, *The Republican War on Science.*

28 Napier and Jost, "Why are Conservatives Happier Than Liberals?"

29 Mondak, *Personality and the Foundations of Political Behavior*; and Hirsch et al., "Compassionate Liberals and Polite Conservatives: Associations of Agreeableness with Political Ideology and Moral Values."

CHAPTER **6**

Different Slates

Listening to people's political views can sound like listening to a reflex . . . it just sounds like something in the wiring.

Colin Firth

All is disgust when a man leaves his own nature and does what is unfit.

Sophocles

On Wednesday, September 13, 1848, a construction gang working for the Rutland and Burlington Railroad was busy blasting rock to clear the way for a new stretch of track just outside Cavendish, Vermont. It was not a job for the faint of heart. It involved boring a hole into a rock bed and filling it with stuff that went bang; you definitely did not want to be standing next to one of those boreholes when it went off. The really dicey part was tamping down the explosives, which required taking an iron bar and ramming down a charge of blasting powder to make sure it was packed firmly. This was a job that had to be done carefully and correctly, and on this particular day, late in the afternoon, the 25-year-old foreman of the work gang biffed it. He jammed his tamping iron down a bit too enthusiastically, detonating the explosive and launching what amounted to a sharpened 13-and-a-half pound metal broom handle into his face. The tamping iron shot into his left cheek, tore through his left temporal lobe, and came out the top of his skull before describing a short arc and landing about 25 yards behind him. The name of this unfortunate fellow was Phineas Gage.

Believe it or not, having his cranium traumatically ventilated did not kill him.[1] In fact, the immediate aftereffects were surprisingly minimal. He spoke lucidly within a few minutes of the accident, walked with virtually no assistance, and proceeded to live another 12 years. Yet Gage did not exactly recover. That sort of injury leaves physical scars, of course, but the story of Phineas Gage continues to fascinate because the injury led to drastic changes in his personality. By all accounts, before the accident Gage was an industrious and upstanding fellow, but afterwards he was described as a moody, depraved, and quarrelsome wastrel (there is something of a debate about whether these changes have been exaggerated, but there is little question that Gage's personality and psychology did indeed change).[2]

Gage's story has been widely popularized by academics as a standard case study of how social attitudes and behavior can change as a result of brain injuries, but Gage is far from the only example of this sort of occurrence. Researchers like the physician and neuroscientist Oliver Sacks (author of *The Man Who Mistook His Wife for a Hat*) write books on how neuroanatomical trauma or abnormality can radically alter social behavior.[3] There are textbooks devoted to explaining the ability of traumatic brain injuries (TBI) to alter neural structure and functioning, thereby affecting mood, personality, and cognitive styles.[4] A depressingly fast-growing research literature documents the psychological effects of TBIs suffered by veterans of the wars in Iraq and Afghanistan.

All of this bluntly and inarguably demonstrates that biology and psychology are inextricably linked; alterations in biology can lead to changes in personality, tastes, preferences, perceptions, attention, emotional experiences, and the attitudes and behaviors motivated by emotions, and that includes attitudes and behaviors pertaining to politics. Our brains are delicate pieces of machinery; perforate an individual's brain bucket in any meaningful way and there is a reasonable chance his or her personality, cognitive functioning, or one of the other critical dimensions of psychology that make up "Jane," "Joe," or "Phineas" will be scrambled. The name and even the physical appearance might remain the same, but Phineas just ain't Phineas anymore.

Doctors have known for centuries that the biological particulars of the brain shape our psychology and thus our behavior, and they have put that knowledge into practice to surgically treat psychological disorders. For example, in the twentieth century roughly 50,000 people in the United States underwent prefrontal lobotomies, a procedure involving surgically slicing through connections to the prefrontal cortex. At one point this was a fairly mainstream treatment for maladies ranging from chronic anxiety to schizophrenia. The main "positive" impact of lobotomies was to lessen anxiety, and people who were lobotomized not only worried less, they seemed to experience less emotion, period. The side effects of poking knitting needles into someone's prefrontal cortex were often devastating. Some patients, such as President John F. Kennedy's sister Rosemary, who was lobotomized in the summer of 1941, were left completely incapacitated.[5] What is interesting for our purposes, though, is not that the brains of lobotomized patients were different after the medical procedure, but that they were almost certainly different *before* surgery. Something in the wiring was causing a set of behavioral or psychological symptoms that was diagnosed as pathological, and the "cure" was to get in there and snip a few of the crossed wires. Doctors who performed lobotomies recognized that the psychological pathologies they were trying to treat (albeit in a brutal and often ethically indefensible manner) were biologically based.

Modern researchers increasingly recognize that underlying differences in brain structure, function, and processing correlate with differences in aspects of psychology that are not pathological. We have different personalities, different tastes and tendencies, different attentional patterns—at least in part because we are biologically different. Numerous observable differences exist in everything from the size or volume of particular brain areas to the intensity with which these different areas are activated in response to environmental stimuli. Behaviorally relevant biological differences associated with tastes and tendencies, though, are not limited to the brain or even to the central nervous system. Numerous aspects of physiology, including hormone levels and subconscious facial expressions, correlate with tastes, preferences, attitudes, and behaviors. Alter the operation of these physiological systems by doing something like shooting an iron bar through your head and, assuming you survive, you might get more than the mother of all headaches—you might wake up quite literally a new you.

At one level, this all may seem obvious. If someone's brain is altered, it seems logical that there will be psychological consequences. Yet the principle Phineas

Gage and the victims of lobotomies exemplify—that differences in biology lead to differences in psychology—applies even to more modest biological variations. Dopamine is an important chemical substance that permits the sending of signals in certain parts of the brain. Subtle differences in the dopaminergic systems of ordinary people can lead to major differences in behavior. We know this because behaviors change markedly when dopamine systems are manipulated artificially. Dopamine-relevant drugs such as pramipexale (sometimes administered to Parkinson's patients) can trigger gambling, overeating, and other addictions. Cut the dosage back and these behaviors will go away. If artificial manipulation of the dopamine system can have these effects, then the substantial naturally occurring variation in this system can have equally telling behavioral consequences. Other behaviorally relevant systems, contrary to the assertions of actor and Scientologist Tom Cruise, respond with equal sensitivity to slight chemical changes. Physiological differences across people in the "nonclinical" part of the population are readily observable and lead to differences in likes and dislikes, and to the types of objects and situations that grab attention. Given what we have just learned about tastes, tendencies, cognitive styles, and politics, might some of those measurable physiological differences also systematically correlate with political orientations and temperament? In this chapter we are going to explore this question in some depth, reviewing a growing body of evidence suggesting the answer is a definitive yes.

I Feel It in My Gut . . . and Maybe a Few Other Places

"The brain recalls just what the muscles grope for," says Rosa Coldfield in William Faulkner's *Absalom, Absalom!* "No more, no less."[6] What Faulkner was having Rosa express with these words is the idea that our minds and our bodies are somehow connected. This intuitive link between the mind and the body is reflected in everyday language. A hunch is something we feel in our gut, unreciprocated love makes our heart ache, sadness leads to a long face, and a teenager who forgets to take out the trash gives us a pain in the ass. This connection between mind and body is also a scientifically validated phenomenon.

The discipline that studies the link between psychological and physiological states is called, with typical academic creativity, psychophysiology, a relatively new science based on the old idea that humans experience the world through their physiology. In some respects, our physiology is simply a series of information processing systems. We experience the world through these systems; they tell us if it's too warm or too cold, if something tastes good or is disgusting, if something is pleasing or disagreeable. Based on this information, these systems adjust physical states to match environments: If it's too hot we sweat; if we are in a pleasing and safe place our muscles relax; and if there is a bear headed our way digestion shuts down and we start pumping out adrenaline to jack up our heart rate and get our lungs pumping. These systems, though, do not merely change our physical states, they also change our psychological states. Physiological changes lead us to feel afraid, attentive, happy, sad, or disgusted.

These psychological states are typically experienced as emotions, and for psychophysiologists emotions are "action dispositions," the motivators or precursors of behavior.[7] Think of it like this. It is a hot day and your sister and her husband come for a visit. Your irritating, fanatical environmentalist brother-in-law marches in and turns off the air conditioning and gives you a lecture on greenhouse gases and global warming. You turn the AC back on. He turns it off. You hit him in the face. Your sister is incensed that you have given her husband a shiner and stamps out vowing never to speak to you again. What is obvious in this domestic fracas is a series of emotion-motivating actions. Your brother-in-law feels disgusted that you so selfishly ignore the health of the planet. The pompous twit makes you feel angry. Your sister is anxious and stressed by the friction between you and her husband. In social situations, the actions of one individual trigger a psychological state—an emotion or feeling—in another, motivating that individual to take an action. Note that it motivates or predisposes; it does not determine. We did not have to punch our brother-in-law; we could have counted to 10, overridden that impulse, and accepted his views. Instead, we just knew that hitting him in the face was going to feel sweet. Emotion, once more, predisposes behavior.

All these emotions are accompanied by physiological states. Anger activates your fight or flight system, which, largely outside of conscious control or awareness, physically prepares you for combat, dilating your pupils, increasing your heart rate, and pushing nutrients towards your muscles. Your sister's stress activates her hypothalamic-pituitary-adrenal (HPA) axis, boosting her levels of cortisol (the "stress" hormone). As your fist closes in on your brother-in-law's mug he reflexively shuts his eyes, pulls his head down, and contracts his muscles into a defensive posture. What all this shows is that physiological and psychological states are connected, and this is the essence of the field of psychophysiology: The mind has a literal physical substrate. Measure that substrate and you can acquire information about people's states of mind, even if they do not want you to know and even if they are not consciously aware of their own state of mind.

To many people, the basic notion of gaining access to psychological states by measuring physiology is passingly familiar thanks to the polygraph, the most common form of lie detector. The central assumption of a polygraph is that someone who is fibbing will feel nervous or guilty, even if those feelings are buried deep in their subconscious. The belief is that those feelings can be detected because they will trigger sub-threshold but measurable physiological changes that give away the falsehood, such as telltale shifts in blood pressure, heart rate, respiration, and electrodermal activity. Physiological functioning, though, has the potential to do much more than help law enforcement interrogators finger prevaricating crooks. It can help to identify the elements of the environment that trip our triggers, even if we have no conscious awareness that our triggers are being tripped. The frequently experienced sense that something "just feels right" (or wrong) is not merely a mystical muse whispering directions in our ear; it is our biochemical HVAC system adjusting our psychological thermostats.

The particular biological mechanisms we are talking about here are our nervous systems, and psychophysiologists tap into the activity of the nervous system by measuring the organs it controls. It is possible to measure this activity because the nervous system is primarily made up of a set of specialized cells called neurons, with big assists from glands and muscles. When these systems fire, the electrochemical signals generated can be picked up by sensors that psychophysiologists stick on the body to read the activity of everything from brains to hearts and muscles. Those signals can reveal a good deal about someone's psychology and they constitute a promising approach for investigating the biological basis of political temperament.

The human nervous system is complicated[8] and includes sensory systems, which rely on specialized neurons to detect things like light, pressure, temperature, and motor systems, which make possible everything from locomotion to nose picking. We will focus on two portions of the nervous system. The first is the central nervous system, or CNS, which technically consists of all the neurons and organs enclosed by bone—basically those that are between our ears or running through our spinal column. The second is the autonomic nervous system, or ANS, which can be thought of as the "regulator, adjuster and coordinator of vital visceral activities."[9] The ANS has two interconnected subdivisions, the sympathetic nervous system (SNS) and the parasympathetic nervous system (PNS). The SNS can be roughly thought of as our "fight or flight" system; its job is to mobilize the body to respond to threat, stress, or arousal. Activate the SNS, for example, and heart rate increases and blood is directed to voluntary muscles so we can use them to run from a bear or walk across the room to embrace a loved one. The PNS is the laid-back counterpart to the SNS; if the latter is the fight or flight system, the former can be thought of as the "rest and digest" system. Activate the PNS and heart rate slows down and blood is shunted off to less conscious activities like digestion. In this chapter, we are going to talk more about the SNS, not because the PNS is unimportant, but because it is easier to measure.

Physiological characteristics tend both to be stable over time for any given individual and also to vary substantially from individual to individual. People who are the jumpiest when startled tend to be that way consistently, just as those who are laid back tend to be so persistently. In a given group, those individuals highest in certain physiological traits at one time tend also to be among the highest in the group at a later time. Our lab has tested the same subjects years apart and found that, barring major changes in health, medications, and the like, their baseline physiological states as well as their responses to general categories of stimuli are quite similar. Dark rooms and loud noises will cause some people's heart rates to go through the roof. Give others the same package of stimuli and their cardiac response is undetectable. More to the point, we have found that these relatively stable individual differences in physiology consistently correlate with differences in political temperament. And, as it turns out, we are far from the only researchers to draw this conclusion.

Politics on and in the Brain

Colin Firth has received one of the rarest accolades ever achieved by a Hollywood A-list movie star. No, we don't mean his Oscar or his Golden Globe. Firth is, as far as we know, the only Tinseltown titan who has published in *Current*

Biology, a big-deal academic journal.[10] In faculty lounges, that is the sort of accomplishment that gets tongues wagging to a much greater extent than an award-winning turn as a stuttering monarch in *The King's Speech*. So just how did Firth come to be listed as a coauthor on a study investigating differences in, of all things, neuroanatomy? Well, it had to do with his unashamedly liberal politics. As he put it, no doubt only partially tongue in cheek, "I just decided to find out what was biologically wrong with people who don't agree with me."[11] To his own astonishment, the scientists Firth commissioned to do the study actually came up with an answer.

The study in question, which was conducted by Ryota Kanai and Geraint Rees at University College London, did not actually discover anything biologically wrong with people holding certain political beliefs, but it did find that liberals and conservatives are, in fact, biologically different. To explore the topic, the researchers first asked 90 young adults living in greater London area to report their political views on a standard five-point "very liberal" to "very conservative" scale. Then they popped these people into a magnetic resonance imager (MRI). This is basically a tube ringed with powerful magnets that are used to line up the atomic nuclei in soft tissue cells like so many iron filings. Mix in some physics and math to this technical wizardry and the end result is the capability to take a picture of structures inside the body. The particular bodily features of interest to Kanai and Rees are known as the anterior cingulate cortex (ACC) and the amygdala. From the side, brain schematics of the ACC make it look a bit like a banana folded into the mid-front of our brains. In similar pictures, the amygdala looks like a couple of almonds hanging out deeper inside.

They had good reason for focusing on these two particular brain regions. Like most parts of the brain, the ACC is a bit of a multitasker, but a couple of things that consistently seem to activate it are tasks involving error detection and conflict resolution. These regions are interesting for those seeking biological correlates of political temperament because they are also associated with the particular patterns of thinking—or cognitive styles—that distinguish liberals from conservatives. For example, in the previous chapter we talked about a study by Russell Fazio and Natalie Shook in which liberals and conservatives played the computer game BeanFest. In that game the cognitive style of liberals was characterized by a higher tolerance for ambiguity and novelty. Basically, they had a "hey, a new bean, let's check it out" approach. Conservatives were much more structured and persistent. They had more of a "recognize the good bean, pick it; never seen the bean, avoid it" style of play.

Researchers figured that these sorts of cognitive styles likely manifested themselves in the organ that makes cognition possible; that is, the brain. This was actually demonstrated years ago by a group of researchers using a technology called electroencephalography (EEG), which consists of participants putting on what looks like a hairnet dotted with sensors that pick up the electrical activity of neurons firing in the brain. In layman's terms, it measures brain waves. EEG technology is attractive to neuroscientists because it has the ability to measure the timing of brain activity more accurately than fMRIs, and it can do so in response to a laboratory-controlled stimulus (these are known as event related potentials, or ERPs). In addition to timing, by identifying the sensors that are picking up these changes, researchers can get a rough idea of the source.

This research method was employed by a team of researchers led by David Amodio, a psychologist at New York University, who in 2007 conducted a study that involved giving a set of subjects a basic go/no-go task while recording their brain waves. The go/no-go task is pretty much what it sounds like; subjects are told to hit one computer key when a specified "go" stimulus is displayed and to refrain from hitting that key when another stimulus appears. Subjects are presented with numerous consecutive "go" stimuli so that they become habituated to pushing the corresponding key. Once in a while, though, a "no-go" stimulus appears, thereby creating a conflict. Subjects are poised to push the "go" key but now a correct response requires the brain to override the ingrained response so that no action is taken.[12]

The basic purpose of the Amodio study was to figure out whether political orientation correlated with neurocognitive activity in the ACC in this sort of error detection/conflict scenario. They measured a specific brain wave component that is associated with error detection and found that the amplitude of this wave correlates with self-reported ideology (extremely liberal to extremely conservative) of their subjects. Essentially, for those on the left of the political spectrum, the ACC sparked hard when the unfamiliar no-go stimulus popped up; the response was less for those on the political right. The key conclusion of this study is that neurocognitive sensitivity to the sort of internal conflict created by the go/no-go task varies according to political ideology.

So if an Oscar-winning actor wants to know where his brain might be different from the brains of those with whom he disagrees politically, previous research suggests the ACC might be a good place to start. The Firth-funded team found out that the ACCs of liberals are not only whizzing the electric meter at double speed under certain conditions but, compared to conservatives, are physically different. Among their 90 subjects, Kanai and Rees found a strong relationship between the volume of gray matter making up the ACC and political orientation. Indeed, the correlation between ideology and ACC size was –0.27, a respectable number for the sort of relationships we are investigating. The more conservative (coded higher on the ideology scale) you are, the smaller the ACC. Liberals, in other words, have bigger bananas.

That does not mean that conservatives should have brain banana envy. At least in Kanai and Rees's subjects, liberals might have had the fruit, but conservatives got the nuts. Just as Kanai and Rees did not alight on the ACC by chance, they also had good reason for looking at the amygdala as a potential discriminator of political temperament. Like the ACC, the amygdala is another multitasking bit of brain. It is typically considered part of the limbic system, a set of brain structures that plays an important role in regulating emotions. Part of the amygdala's role seems to be to help govern attention to emotions. It also plays an important role in social cognition, or the way information about other people is processed, stored, and employed. For example, several brain-scanning studies have found that the amygdala is involved in helping individuals evaluate faces, including making judgments of whether or not a face is trustworthy.[13]

This is interesting in the context of the studies discussed in the last chapter that found conservatives to be more likely than liberals to pay attention to certain types of faces—especially threatening or angry faces. We also described studies that consistently found differences in attentional patterns, with the general theme being that conservatives are quicker to find and to pay attention to negative stimuli. If this consistent set of patterns has a biological basis, then the amygdala certainly seems like a good place to look for it. Kanai and Rees's scans revealed a relationship between political orientation and amygdala volume that was very similar to that between political orientation and ACC volume. As expected, though, the relationship was reversed. The correlation between amygdala volume and political orientation was +0.23; the more conservative the subject the greater the volume of the amygdala.

A note of caution is in order here. Neuroscience has made impressive leaps in knowledge over the past decade or two and greater access to the technology and the experts who run it is allowing an increasing number of social scientists to use these tools to pursue fascinating topics. Still, it is important to keep in mind that there is a good deal that we do not know about the ways in which the structure and processing of the brain connect to cognitive patterns.[14] Some studies seem to suggest that we can flick on a brain imager and literally see what someone is thinking. For example, one study done on 20 voters in advance of the 2008 presidential elections used brain-scan data to draw conclusions about which primary candidates elicited more empathy from the subjects and which party "brand" was evoking more anxiety.[15] That study was strongly criticized by a group of cognitive neuroscientists who bemoaned publication of research without peer review (this particular study was reported in the op-ed pages of *The New York Times*) as well as the tendency to offer misleading impressions of the sort of things that brain scans can or cannot do.[16]

Part of the problem is that the techniques and technologies being used in this area of research are far from perfect. For example, an increasing number of studies employ functional magnetic resonance imaging (fMRI), a technique that allows researchers to measure neural activity, not just neural structure. Extracting and, especially, interpreting this information involves not just science and complicated statistics but also a goodly dose of judgmental art. One of the problems attending fMRI is the risk of false positives; in layman's terms, an fMRI scan generates so many readings (because there are so many "voxels," or small sections, in the brain to analyze) that by chance alone some of them are bound to correlate with whatever is being studied. One example of this danger was provided by a group of psychologists who put a "post-mortem Atlantic salmon"—i.e., a dead fish—in the scanner and showed it a series of pictures depicting scenes of social inclusion or exclusion. The salmon was asked to identify the emotion experienced when viewing these pictures. Sure enough, the researchers managed to get a reading of neural activity that lined up with the hypothesis that a recently deceased fish was taking some sort of perspective on social situations.[17]

While cautionary perspective is certainly advised for any research technique, the oft-noted salmon result is hardly damning of the brain-imaging enterprise. The core message is simply that if you look everywhere you are quite likely to find something. Scholars who, unlike the tongue-in-cheek salmon researchers, go to the work of specifying ahead of time precisely where in the brain a response is expected (these are called "region of interest" studies) will generate findings deserving of confidence. Too many independent fMRI studies have shown that, to take just one example, the amygdala is activated by emotional judgments and threatening stimuli for the value of the technique to be denied.

The take-home point is that the sorts of things we have been cataloging over the past two chapters—the consistent differences in tastes and tendencies and the repeatedly demonstrated distinctive patterns in cognitive style—in an increasing number of studies are lining up, with observable differences in neuroarchitecture and functioning in a way that makes sense. Indeed, we are at the point where researchers can do a better job of predicting whether someone is a Democrat or a Republican by using their brain activation patterns (in response to a "risk" game) rather than the environmental and socialization patterns that dominate political science theories of partisanship.[18] A sensible conclusion is that when it comes to conservative-liberal brain differences, though the specifics are still being worked out, those differences exist and are consistent with findings on the correlation of political orientations with personalities, directed attention, and cognitive biases.

Politics Makes Me Sweat

Biological differences between liberals and conservative are not confined to the mysterious grey goop electrically pulsating in our craniums. They also can be found in the more mundane aspects of our biology that are observable without Star Trek levels of technology. Take sweat. Sweat, or at least the process that produces wet pits, clammy palms, and attendant social awkwardness, can be interesting. Psychophysiologists have been fascinated with the whole phenomenon for decades and have devoted considerable time and energy to the process of perspiration.

What makes sweating so interesting is where it happens: the skin. Neuroscience might be the glamour field for investigating the biological basis of social attitudes and behavior but skin is also a sensory and processing organ, and a big one at that. Of the 78 organs in the human body, skin is by far the largest, accounting for 20 to 25 pounds of a person's total body weight. By way of comparison, a typical brain weighs in at a comparative paltry 3 pounds or so. Your skin is not just a barrier to keep germs out and is not just a nice soft sheath that wraps you in a socially presentable package. Skin is a "giant receptor separating us from the rest of the world."[19] It helps to mediate our interaction with the environment. It is responsive to a variety of signals that originate within and outside our bodies. Those signals can be picked up by measuring changes in the electrical properties of skin, which have long been known to fluctuate based on internal psychological states (these fluctuations are generically known as electrodermal activity, or EDA). Compared to electrical activity in the brain, measuring the skin's electrical properties is relatively straightforward and can now be done with an extremely high degree of accuracy.

This accuracy has not provided a foolproof way to figure out if a given individual is telling the truth or telling tall tales. Polygraphs, after all, can be beaten or can damn an innocent as a liar. Yet while EDA does not give us a sure means of figuring out if Colonel Mustard committed the murder in the library with a candlestick, it can tell us something important about general patterns of (autonomic) nervous system activity. We know this because, unlike some of the mysterious processes of the brain, scientists have a pretty good grasp on what makes us sweat. Sweating is the result of the skin responding to signals coming from the sympathetic nervous system. The SNS, remember, is responsible for preparing the body for action. Part of this preparation entails opening sweat glands. When the SNS activates these glands, moisture is drawn to the surface of the skin, sort of like racks of teeny straws sucking up fluid. A hot room or the thought of a hot date can both induce sweat, but only the former involves ambient temperature. The latter is an example of the SNS responding to an internal psychological rather than an external environmental state. It is responses to those internal signals that most interest us, and we can investigate the topic pretty effectively because one type of sweat gland, the eccrine gland, is particularly sensitive to SNS signals generated by internal psychological states.[20] Eccrine glands are densely concentrated in the palms of our hands, and the key thing to remember is that when the SNS perks up because of an internal psychological state like fear or arousal, thousands of those little straws in and around our palms start sucking moisture toward the surface of the skin.

SNS activity can be inferred by the electrical properties of the skin. Just as the most efficient way to get an electric current from one end of a bathtub to the other is to fill it with water, skin in the vicinity of open eccrine glands will conduct electricity faster than skin with less moisture. Those little straws can be thought of as a dense array of electrical resistors controlled by the SNS; as moisture moves up and down inside them they regulate how efficiently the skin can conduct an electrical current. That fluctuation in conductance or resistance (EDA) directly measures SNS activation or deactivation. This well-understood phenomenon makes EDA a simple and direct means of measuring SNS activation and "one of the most widely used . . . response systems in the history of psychophysiology."[21] A simple way to measure EDA is to run a (very) small electrical current between two sensors on the fingertips or the palm and measure changes in that current. (This is what we do in our lab.) If the current spikes in response to a particular stimulus, it is safe to conclude that the stimulus is jacking up the subject's SNS. If the current drops, the SNS is gearing down.

Earlier we referred to the SNS as the body's "fight or flight" system. This is accurate as far as it goes. The SNS prepares us to run from a bear or wade into a smackdown with our brother-in-law. Still, fight or flight is an incomplete description of what the SNS does. The SNS perks up when we need to pay attention or think hard. It mobilizes resources that we might need when we have to take any sort of action, be it action in response to a suspiciously bear-shaped shadow in the berry bush, a tenth-grade algebra problem, or a loved one. And don't forget that the SNS operates partially, though not entirely, outside of conscious awareness. We might feel our heart racing if we see a bear close by, but physiological changes in response to less dramatic events are often outside of conscious awareness. We have measured people's physiological response to a visual stimulus on a computer screen and then asked them to rate the intensity of their response and found that the self-report and physiological measures do not match well at all. Equally true is that if a person's SNS makes him or her particularly sensitive to bear-shaped shadows in the bushes, that individual is more likely to play it safe, stay out of the bushes, and gear up to do an Usain Bolt impression. Another person's SNS might sound less of an alarm bell—it is just a shadow, after all—and would be less likely to be dissuaded from going into the bushes after berries. Those predispositions and the subsequent behaviors they shape can have important consequences. One person has no berries but at least is safe. The other either has berries or is bear food. Regardless, it is apparent that attitudes and behaviors vis-à-vis the shadow are at least partially driven by biologically based predispositions rooted in the sensitivity of the SNS to particular stimuli.

In sum, if political orientations are biologically based, stable individual-level differences in SNS activation seem like another good place to look for them, and indeed a number of studies have found that SNS activation correlates with particular sets of political attitudes and behaviors. One of these studies was conducted by our lab back in 2008. We brought in about 50 adults and showed them several images on a computer screen. Three of these images were rated by independent observers as particularly threatening: a large, hairy spider crawling across someone's face; an open wound with maggots crawling in it; and a dazed, beaten, and bloody man. We measured EDA response to these images and found that it was systematically correlated with a particular set of policy positions. We termed these "socially protective policies" because that is exactly what they seemed to reflect: policies designed to protect the interest of the participant as well as the participant's social group from threats. These issues included the death penalty, immigration, foreign aid, and gun control. We found that individuals with higher EDA response to the threatening images—read higher SNS activation—consistently had more conservative policy stands on socially protective policies.[22]

We took this as supporting the notion that individual-level differences in physiology encourage the adoption of particular political attitudes. In this particular case, people who were more physiologically responsive to threat stimuli were more likely to support policies aimed at reducing, or at least addressing, threats to the social status quo. Whether for or against, people's positions on these sorts of policies often just "seem obvious common sense," but people are notoriously bad at being able to articulate the real reasons for their political attitudes. The physiological results suggest that in actuality they may hold those positions because they simply "feel right."

We put this general idea to a second test in a study where we examined EDA activity in response to disgusting images. Disgust is a particularly interesting emotion to study. In Chapter 4 we discussed differences in tastes and preferences and pointed out that people generally avoid things they find disgusting. Moreover, if you remember the discussion of moral foundations theory, you might recall that conservatives are more likely to emphasize purity and disgust as a foundation for moral and political orientations. Researchers have known for some time that self-reported disgust sensitivity, not to mention the kinds of things found to be socially or morally disgusting, are related to political beliefs such that those who report higher disgust sensitivity are more likely to adopt conservative positions, especially on sex-related issues like gay marriage.[23]

How do we get from disgust sensitivity to support for gay marriage? Well, let's start by recognizing that disgust is a very powerful feeling, "the most visceral of all emotions."[24] If you have ever gagged after smelling or seeing something particularly vile, you know this to be true. Good evolutionary reasons exist for disgust being such a dominant action predisposition. Feelings of disgust lead people to avoid the sources of pestilence—that is what the gag reflex is all about. If a rotting carcass makes us nauseous rather than hungry, we are less likely to eat it and thus more likely to stay alive. Disgust, though, is not limited to avoiding rotting meat or to giving steaming piles of poop a wide berth. This emotion, along with its powerful impact on attitudes and behaviors, transfers to the social aspects of the environment. Most people in most cultures, for example, find the thought of incest disgusting. That makes good evolutionary sense. If Oedipal thoughts make you feel queasy, you are less likely to end up with jug-eared kids playing one-stringed banjos. Evolution, in short, seems to have used disgust as a means to avoid the fitness cost accompanying the practice of close genetic relatives producing offspring. It doesn't stop there. Disgust, as we have already hinted, also plays a role in making moral judgments. Some social actions—betrayal and support of out-groups—can induce disgust. This is the general idea conveyed by the quote from Sophocles at the beginning of this chapter, and when we find something disgusting, be it in the domain of microbes, mating, or morality, we seek to avoid and/or condemn it.

Like pretty much everything else, though, there are individual differences in disgust sensitivity in all three of these domains.[25] Just as some people have a gag reflex to foods that others wolf down, it is not surprising that social behaviors or political positions passionately supported by some people tend to make others feel queasy. What is interesting here, however, is that we are not just talking about someone who finds the thought of a same-sex couple getting it on disgusting but rather that opposition to gay marriage is higher in people who are more disgust sensitive regardless of whether that sensitivity is triggered by stimuli connected to microbes, mating, or morality. Think of it like this: People who are more disgust sensitive logically seem to be more likely to take disgust/purity concerns into account when making a moral (read political) decision. Given the logic of moral foundations theory explored in Chapter 4, it follows that disgust sensitivity should make people more likely to be conservative, particularly on issues that combine mating with morality—topics like abortion and gay marriage.

Disgust has a well-known physiological signature and part of that signature is activation of the SNS, which, as we have seen, can be picked up with EDA measurements.[26] In our study, we showed people some really disgusting pictures—e.g., poop in a toilet and a person eating worms. Decorum and the need to keep laboratory images out of public circulation means we cannot show you the actual images, but Figure 6.1 gives you a general idea of the type of images seen by the subjects. We measured changes in EDA in response to viewing these pictures. We then compared the degree of change with the subject's positions on a range of political attitudes and discovered a correlation of about .30 between this collection of political attitudes and these physiological changes, suggesting that EDA increases are greater for conservatives than liberals. That general conclusion is a bit misleading, though. When we broke down the correlations by specific attitudes, only a handful—all sex related—were really driving things. Conservative positions on taxes and size of government appear to be little affected by disgust responses, but the correlation with many social attitudes was high and the correlation with gay-marriage attitudes was 0.44; in other words, the relationship between physiological disgust response and this big contemporary morality/mating political issue was huge.[27] The take-home point of this study, though, is not just that people who are more physiologically responsive to disgusting stimuli tend to adopt conservative positions on sex-related policy issues such as gay marriage and abortion. More important is that these findings offer another piece of evidence that individual-level variation in physiology predisposes people to adopt particular political positions.

We are not the only people to use EDA to make this general point. For example, policies that provide some sort of preference to racial minorities, or grant a specific recognition of the rights of such groups (sometimes called affirmative action), have been bitterly contested in American politics since the 1960s. Similar controversies have erupted in Europe as racially and religiously diverse waves of immigrants have splashed different hues and cultures across the traditionally off-white demographics of many nations.

Opposition to such policies could be grounded in honorable and defensible political principles. If you believe that the government should treat all groups equally but the government instead has a set of policies that you perceive as not doing that—singling out certain groups for special recognition or benefits—then you have a perfectly legitimate political beef. Dealing with that beef by asking government to treat all races equally does not by any stretch of the imagination mean your political attitudes are racist.[28]

On the other hand, opposition to affirmative action could also spring from racism. This matter is challenging to resolve since, apart from the occasional neo-Nazi or xenophobe, people with brazenly racist prejudices recognize these attitudes are not socially acceptable and thus are unlikely to admit to them. Such individuals probably would point to principled conservatism as a defense for their policy attitudes rather than fess up to being racists. Determining the real source of attitudes on affirmative action is nigh on impossible using standard

Figure 6.1   Example of a Disgusting Image
This is the sort of image we showed research subjects so we could measure how they responded to disgusting images. This is actually one of the authors, which just shows what we're willing to do in the name of science.

survey approaches. A potentially valuable alternative is EDA. It has long been known that people react physiologically to the presence of out-groups. Is it possible that people who have stronger physiological reactions to out-groups might also be more likely to oppose affirmative action sorts of policies, even if they are not truly sure why? That the SNS will activate in the presence of politically relevant out-groups is not really in dispute. One of the first ever demonstrations of this fact is directly relevant to our affirmative action example; more than 50 years ago, researchers noticed that the EDA of white subjects jumped if they were dealing with black lab proctors.[29] This finding set off an ongoing research agenda that is aimed at uncovering the physiological correlates of racial attitudes.

A more recent study was conducted along these lines by a research team at the Université Blaise Pascal in France. The team showed a set of mostly white French college students pictures of two people: "Sebastien," a stereotypically French fellow, and "Rachid," a stereotypically Arab fellow. The students were asked to rate these two on the basis of their "likeability" and "cleverness." These survey items showed no statistically significant differences; the college students were reporting that they thought Sebastien and Rachid seemed equally likeable. EDA responses to these pictures, though, told something of a different story. EDA response was greater (SNS activity was higher) when the students were looking at Rachid. These subjects claimed to be judging these two guys as equals; they may have even believed that. Yet their fight or flight system was kicking in harder when they were looking at Rachid.[30]

Of course, physiological reactions do not have to drive attitudes, but there is at least some evidence that they can. Researchers at Emory University asked a group of students to review a set of applications for prestigious fellowships, and these included pictures (two white, one black). The selections were supposedly made on the basis of merit, but the students who physiologically reacted more strongly to pictures of out-group members (i.e., blacks) were more likely to pick white applicants as more deserving of the fellowship.[31] These subjects may have genuinely believed that their fellowship recommendations had nothing to do with the race of the applicants, but activation of the SNS is noticeable—and in this case it is not just connecting to an attitude but to a decision.

All of these studies support a key point from our discussion of brain studies: Attitudes and behaviors that are indirectly or directly political have biological correlates. We all have different information processing systems that are apparent in traits of our nervous systems. The typically outside-of-conscious-awareness assessments of the autonomic nervous system clearly seem to make certain political attitudes and behaviors more appealing to some than others.

In Your Face Politics

Imagine that a random social scientist asks you to participate in a study and you agree. To your disappointment, you are not given 20 dollars, as was the case in Chapter 3, but 20 photographs. They are generic black and white portraits of white males and show only their faces; not their hair or how they dress, just their mugs. The social scientist asks you to sort the photos into two piles—one for men who are conservatives and one for men who are liberals. Could you do this with any degree of accuracy?

At first blush, systematically identifying political orientation from a quick look at a face seems impossible. Even if we assume that all the males in the pictures are actually either a liberal or a conservative and not something in between, we still seem to have only a fifty-fifty shot at correctly identifying their ideology. Without further information it seems likely our two stacks are going to reflect little more than guessing. Yet a number of studies suggest otherwise. People turn out to be remarkably good at identifying political orientation just by looking a person's face.

This was first demonstrated in a 1954 study in which subjects were asked to create one pile of British Labour Party supporters and one of Conservative Party supporters. Unbeknownst to the subjects, what they were sorting were generic portraits of back-bench members of Parliament, and they correctly identified party loyalties significantly more than chance would suggest.[32] This seemingly mystical ability of people to identify accurately supporters of the two major political parties in British politics has been replicated several times in the intervening decades.[33] And it is not just the Brits who somehow seem to wear their politics on their faces. One recent study done in the United States found people could not only identify the party membership of legislative candidates using nothing more than facial information, they could also pretty accurately identify a person's party membership from high school yearbook pictures.[34] This is not a case of picking up some decoder ring cultural clue that has yet to be understood, since people can quite accurately identify the ideological lean-ings of people from other countries as well as their own.[35]

Why are faces so revealing? If political temperament is biologically based, it makes sense that it is being broadcast by faces. Our faces are constantly, with-out any conscious input or even awareness, beaming to the world information about our feelings and social intent. Faces are the visual Twitter accounts of our nervous systems, able to distribute information about psychological states quickly and succinctly, and to many people at the same time. At least since Darwin, researchers have recognized that the face provides a universal means of human social communication. We can quickly and accurately assess someone's psychological state—whether they are happy, sad, ticked off, surprised—with a glance at his or her face. This form of social communication is so fundamental to human nature that psychophysiologists argue that "without [facial expres-sions] individuals do not communicate, do not affiliate, do not proliferate, do not interact—in short, are not social."[36] Indeed, faces are said to "leak" our internal psychological states; we involuntary smile, frown, or wrinkle our noses when we feel joy, disapproval, or disgust. Humans may not all speak the same tongue, but we are universally fluent in face. We are also pretty good at detect-ing facial fibbing. We can usually tell, for example, when a smile expresses real joy or is just being faked for social consumption.[37] Of course, there are people who are really good at faking it (actors, for example), but most people find fak-ing impossible to do convincingly.

Faces do more than provide a way to communicate our feelings about our brother-in-law. They declare membership in socially meaningful groups. Some of this is intuitive. Faces, for example, make it easy to classify someone into a particular gender, racial, or age group. Studies also show that people can accu-rately predict an individual's sexual orientation and even religious affiliation using only facial information. They are able to do this after looking at a face for just a fraction of a second.[38] And, as already discussed, the declaration of social affiliations apparent from our faces also includes political orientation.

Some of these studies suggest that people divine political orientation from faces by perceiving them as more or less powerful or socially superior. At least one study finds people think conservative faces look more intelligent.[39] These are purely perceptual judgments, though, and do not actually measure any-thing about the face. We were interested in exactly what it is physiologically about faces that signals political orientation, and we suspected it might have something to do with the degree of emotional expressivity in a face. One of the aspects of personality known to separate liberals and conservatives (or at least partisan affiliations) is expressivity. For example, two psychologists, James Gross and Oliver John, developed a sort of personality test called the "Berkeley Expressivity Questionnaire" that is designed to measure individual-level variation in emotional expressivity. Democrats tend to score higher than Republicans on this set of items.[40]

If facial expressions are known to be a primary and largely sub-threshold means of signaling emotional states, then it follows that Democrats (liberals) will also tend to have relatively expressive faces. We tested this hypothesis using a technique known as electromyography (EMG), a fancy term for putting sensors on the skin to measure the electrical activity picked up by muscle contractions. The specific muscle measured was the *corrugator supercilii*, found between the eyebrows. Its job is to furrow the brow. Even if we are not aware the muscle has moved, negative emotions like disgust, anger, and fear tend to activate the corrugator; positive emotions tend to make it relax. Corrugator activation or deactivation helps to create the facial expressions associated with many of our primary emotions. Accordingly, we prevailed upon a group of adult subjects to tell us their ideological leanings and later we measured their facial expressivity by the extent of their corrugator activity in response to a variety of positive and negative stimuli. Our hypothesis was that liberals would be more facially expressive than conservatives and that turned out to be half true. Like other EMG studies, we found females, regardless of political persuasion, to be more facially expressive than males. Unlike any other EMG study, we found liberal males to be emotionally expressive at pretty much the same level as females. The most distinctive group by far was conservative males. While corrugator activation in response to the images was significant for everyone else, for conservative males it didn't budge.[41]

Perhaps people are able to discern personality traits and therefore political orientations from images (most of the studies use pictures of males) because stoic, less expressive faces (think Clint Eastwood) signal traits associated with conservatism and sensitive, more expressive faces (think Alan Alda) signal traits associated with liberalism. Certainly these signals are not 100 percent accurate, but they do permit quick judgments that appear to be right more often than they are wrong. This conclusion is supported by another study that took "liberal" and "conservative" faces and created avatars that exaggerated facial features and expressions. The liberal avatar was smiling, with a relaxed corrugator; the conservative avatar had less of a smile and even looked a bit frowny.[42] Evidence that political temperaments are instantiated in our biology is found not just in individual-level variation in our brains or the internal wiring of our autonomic nervous systems. Quite literally, politics is also on our faces.

Conclusion: Physiopolitics

A growing body of evidence documents that political temperaments have biological substrates. We have focused mainly on the central and autonomic nervous systems, but other aspects of physiology also correlate with politics. For example, various studies have linked hormones like cortisol, testosterone, and serotonin to political attitudes and behavior (though not, as far as we are aware, to ideology).[43] Other studies even suggest that muscle mass correlates with political attitudes. At least among males, the more buff you are, the more likely you are to push strongly for positions that further your own economic interest (socialistic redistribution if you are poor; laissez-faire capitalism if you are rich).[44]

It is true that more replication is needed before complete confidence can be vested in the links between biological processes and structures and specific dimensions of political temperament. Empirical research on the connection of biology to politics is in its infancy. Still, there is simply too much evidence from too many sources to credibly argue that political attitudes and behaviors have no connection to biology. Beyond establishing this link, what does the new research tell us about differences between liberals and conservatives? People have quite different nervous systems. Some more than others have sympathetic nervous systems primed to respond more strongly to particular stimuli. Combine these differences with those in the CNS and the rest of the ANS, and toss in endocrine systems for good measure, and the end result is that people physically experience the world differently. One person may look out and see threats—look at the bear shadow! Another may look at the same view and see opportunity—look at the berries! These different perceptions are based in physiological responses and will unavoidably affect the manner in which people operate in their social and political worlds. The extent of negativity bias (perceiving, responding, and attending to aversive situations more than pleasant situations) varies from person to person and consistently is higher for conservatives.[45] People who support greater military spending, harsher punishment for criminals, and restrictive immigration are not doing so just to infuriate liberals but because they are more physiologically and psychologically attuned to negative eventualities. The next question is whether these deep-seated differences are attributable to short-term environmental forces or run deeper, all the way to our DNA.

Notes

[1] Macmillan, "Phineas Gage—Unraveling the Myth."
[2] Ibid.
[3] Sacks, *The Man Who Mistook His Wife for a Hat.*
[4] Silver et al., *Textbook of Traumatic Brain Injury.*
[5] National Public Radio, "Frequently Asked Questions about Lobotomies." Available at http://www.npr.org/templates/story/story.php?storyId = 5014565.
[6] Faulkner, *Absalom, Absalom!*
[7] Lang, "The Emotion Probe: Studies of Motivation and Attention."
[8] A good introductory text on psychophysiology that includes an accessible discussion of the underlying biology of the human nervous system is Stern et al., *Psychophysiological Recording.*
[9] Noback and Demarest, *The Human Nervous System: Basic Principles of Neurobiology.*
[10] Kanai et al., "Political Orientations Are Correlated with Brain Structure in Young Adults."
[11] BBC Radio Four, "Colin Firth: An Opportunity to Explore." Available at http://news.bbc.co.uk/today/hi/today/newsid_9323000/9323470.stm.
[12] Amodio et al., "Neurocognitive Correlates of Liberalism and Conservatism."
[13] Rule et al., "Face Value: Amygdala Response Reflects the Validity of First Impressions."
[14] Westen, *The Political Brain.*
[15] Iacoboni et al., "This Is Your Brain on Politics."
[16] Aaron et al., "Politics and the Brain."
[17] Bennett et al., "Neural Correlates of Interspecies Perspective Taking in the Post-Mortem Atlantic Salmon: An Argument for Proper Multiple Comparisons Correction." The authors won a 2012 Ig-Nobel Prize for this research!
[18] Schreiber et al., "Red Brain, Blue Brain: Evaluative Processes Differ in Democrats and Republicans."
[19] Stern et al., *Psychophysiological Recording,* 206–207.
[20] Ibid., 209.
[21] Dawson et al., "The Electrodermal System."
[22] Oxley et al., "Political Attitudes Vary with Physiological Traits." Preliminary evidence from this same study also suggests that individuals supportive of socially protective policies tend to have an elevated "startle" response—in other words, their muscles moved significantly more in response to an unexpected auditory startle.

23 Haidt and Hersh, "Sexual Morality: The Cultures of Conservatives and Liberals"; and Inbar et al., "Conservatives Are More Easily Disgusted Than Liberals."

24 Harrison et al., "The Embodiment of Emotional Feelings in the Brain."

25 Tybur et al., "Microbes, Mating and Morality: Individual Differences in Three Functional Domains of Disgust."

26 Harrison et al., "The Embodiment of Emotional Feelings in the Brain."

27 Smith et al., "Disgust Sensitivity and the Neurophysiology of Left-Right Political Orientations."

28 Sniderman and Carmines, *Reaching beyond Race*.

29 Rankin and Campbell, "Galvanic Skin Response to Negro and White Experimenters."

30 Dambrun et al., "On the Multifaceted Nature of Prejudice: Psychophysiological Responses to Ingroup and Outgroup Ethnic Stimuli."

31 Vanman et al., "Racial Discrimination by Low-Prejudiced Whites: Facial Movements as Implicit Measures of Attitudes Related to Behavior." We should note that this particular study did not use EDA but rather a different measure of SNS activity.

32 Jahoda, "Political Attitudes and Judgments of Other People."

33 Bull and Hawkes, "Judging Politicians by Their Faces"; and Bull et al., "Evaluation of Politicians' Faces."

34 Rule and Ambady, "Democrats and Republicans Can Be Differentiated from Their Faces."

35 Samochowiec et al., "Political Ideology at Face Value."

36 Cacioppo et al., "Social Psychophysiology: A New Look."

37 Ekman et al., "The Duchenne Smile: Emotional Expression and Brain Physiology II."

38 Rule and Ambady, "Brief Exposures: Male Sexual Orientation Is Accurately Perceived at 50ms"; and Rule et al., "Female Sexual Orientation Is Perceived Accurately, Rapidly, and Automatically from the Face and Its Features."

39 Bull and Hawkes, "Judging Politicians by Their Faces."

40 Gross and John, "Facets of Emotional Expressivity: Three Self-Report Factors and Their Correlates."

41 Jacobs et al., "Carrying Your Heart (and Your Politics) on Your Face: Ideology and Facial Muscle Responses."

42 Roberts et al., "Judging Political Affiliation from Faces of UK MPs."

43 McDermott, "Hormones and Politics"; and Waismel-Manor et al., "When Endocrinology and Democracy Collide: Emotions, Cortisol and Voting at National Elections."

44 Peterson et al., "The Ancestral Logic of Politics: Upper Body Strength Regulates Men's Assertion of Self-Interest over Economic Redistribution."

45 Hibbing, Smith, and Alford, "Differences in Negativity Bias Underlie Variations in Political Ideology."

CHAPTER 7

Politics Right Down to Your DNA

Social scientists have been working night and day
Checking re-checking re-re-checking DNA
After years of research the knowledge they've acquired
Has scientists thinking we might just be hardwired.

Christine Lavin

On May 3, 1953, two healthy girls named Kay Rene Reed and DeeAnn Angell were born at Pioneer Memorial Hospital in Heppner, Oregon. A couple of days later, they were taken home by families that lived close by—the Reeds in the city of Condon and the Angells about 20 miles south in a town called Fossil. For more than 50 years, they led perfectly normal lives and each loved their parents and siblings deeply. Kay Rene grew up, married a cattle rancher, works in a bank, and has children and grandchildren. DeeAnn grew up, married a car salesman, and became a homemaker; the couple later moved to the elk-hunting region of Mead, Washington. Their seemingly conventional life stories, though, were based on a wrenching mistake: Kay Rene and DeeAnn had been accidentally switched in the hospital.

There were suspicions from the start. When nurses handed Marjorie Angell back her one-day-old daughter after taking her out of the room for a washing, she insisted, "This is not my baby." The indignant nurses huffed, "She most certainly is your baby," and that was that. Majorie bonded with the baby and brought her newborn home to be raised with five siblings, including two-year-old twin boys.

Yet the doubts never vanished, especially as DeeAnn grew and did not always fit comfortably with the rest of the Angells. Marjorie whispered her suspicions to her older daughter Juanita and to her friend Iona Robinson. Iona in turn noticed that Juanita bore a striking resemblance to a daughter in another family she knew—Kay Rene Reed.

Kay Rene was raised by Donalda Reed, who also harbored suspicions and mentioned them to her older children. Still, no one did anything about these suspicions until 2008 when Iona, by then in her late eighties and in failing health, decided to take action. After agonizing for months, she called Kay Rene's brother, Bobby Reed, explaining apologetically that she needed to get something off her chest. She told him that Kay Rene was quite likely not his biological sister. Bobby was stunned and not sure what to do. Both his parents and the Angell parents had passed away, so they could not be consulted. He did nothing for nine months but eventually decided to confide in his sisters. Together they made arrangements to meet DeeAnn. Bobby Reed's first thought upon laying eyes on DeeAnn was that "she looked just like my mom." They told Kay Rene of the situation in March of 2009, and she found the news devastating.

DeeAnn and Kay Rene met and began collecting clues from their childhoods about their accidental fates. Kay Rene once had a dust-up with a high school biology teacher who said it was genetically impossible for two blue-eyed parents to have a brown-eyed child. Brown-eyed Kay Rene insisted that must be wrong because her parents both had blue eyes. Growing up, neighborhood kids teased DeeAnn about being the only blonde in a family of brunettes. The truth concerning their biological heritage awaited DNA testing and steps were taken to resolve the matter. It took three weeks for the DNA report to arrive, a time Kay Rene describes as like "waiting for a cancer screening." She wanted very much to be a Reed, to continue being a part of the only family she had known for 55 years.

The results were unequivocal: Kay Rene had zero percent shared genetic heritage with the other Reed siblings and DeeAnn Angell was 99.9 percent related to the Reeds. The scientific confirmation devastated Kay Rene a second time. DeeAnn initially took the news better but soon was an emotional wreck, waking up in tears and feeling as though she had lost her mother all over again.

With the passage of years, the "swisters" became more reflective, and Kay Rene now admits to "being intrigued by the science of this, nature versus nurture." She found it a relief to know why she was the only Reed to have acne, impaired vision, and plain looks; and DeeAnn, as Bobby immediately recognized, was a dead ringer for females in the Reed family. Appearance, however, is only part of the story. Both Kay Rene and DeeAnn display behavioral and temperamental patterns notably similar to their biological families. Kay Rene chews her fingernails constantly, something nobody else in the Reed family does—but Marjorie Angell did. Kay Rene is blunt and terse; her Reed siblings are anything but. She smacks gum just like one of her biological siblings. Dee-Ann Angell describes herself as "a girlie girl." She likes short skirts and high heels and jokes that she would not go to her mail box without her nails and makeup being done. With regard to temperament and taste, DeeAnn concedes to being "nothing like the Angells I was raised with." While growing up she frequently felt as though she did not belong.[1]

That DeeAnn Angell's blonde hair and noticeable looks would more closely match her biological family than her adoptive family probably surprises no one. Height, hair color, eye color, weight, facial shape, and symmetry are all known to be passed along by genetics, a fact that is apparent to the casual eye as well as in the results of systematic scientific studies. But many people have a difficult time accepting that behaviors, comportment, and attitudes are also connected to genetics. The folk wisdom seems to be that having fingernails is influenced by genetics but the habit of chewing them is not.

To a certain extent this belief is understandable. The notion that complex social behaviors might be under genetic influence makes people uncomfortable; it makes it seem as if we are not in conscious control of our actions and motivations. An alternative explanation must exist for DeeAnn and Kay Rene's behavioral resemblances to their biological families, right? Maybe it is just coincidence. Can genetics really influence taste in clothes, gum smacking, nail chewing, and perhaps political preferences? In answering this question, we will begin by describing research conducted on nonhuman animals. Research on other species with close parallels to human tissue structures and nervous system operations can teach us important lessons about our own species. This has certainly happened on the medical front, where animal models continue to offer crucial health-improving insights, and now the same value is being realized in the realm of human behavior. Strong evidence indicates that genetics shape a wide range of social behaviors in honeybees, silver foxes, fruit flies, and voles and, much as humans would love to believe that our behavior is too sophisticated to be influenced by genes, this is hardly the case.

Paternal Voles, Amorous Flies, and the Fox That Became a Dog

More than any other mammal, Mickey Mouse has made rodents seem cute, cuddly, and socially approachable, but voles run a close second to Mickey. These ground-dwelling, hamster-sized balls of brown fur are not just endearing but also illuminating. Unlike Mickey, who, despite courting Minnie for 80 years, has yet to reproduce, voles have taught us several important lessons about parenting. Voles are major, if unwitting, contributors to current understanding of the roots of paternal activity because their behavior varies so much from vole to vole and because their biology has clear parallels to other species, including humans. Some vole dads are active physical presences in the early lives of their offspring, showering them with licks, attention, and affection. Other fathers stay close to their developing offspring but are not nearly as expressive. Still other fathers are not present at all and take off emotionally and geographically soon after the pups' birth. These distinctive behavioral patterns puzzled researchers. What was causing so much variation on a matter as central to survival and reproductive success as paternal involvement in child rearing?

We now know a major reason that vole dads range from Ward Cleavers to Don Juans is genetics. Evidence for the role of genetics comes from scientists' ability to trace the causal pathway that runs from genes to biology to behavior. To understand that pathway (and much of the rest of this chapter), it is important to understand the nature and function of genes, so settle in for a brief genetic primer. A gene is the essential unit of inheritance because particular versions of genes get passed on from one generation to the next. Genes are found in the deoxyribonucleic acid (DNA) that is present in the nuclei of most of a body's cells and that contains the blueprint for the development and functioning of biological organisms. DNA is like a ladder twisted in the shape of a spiral staircase. The rungs of the ladder are made up of nucleotide base pairs (adenine, thymine, cytosine, and guanine) that join together in certain ways (adenine always pairs with thymine, cytosine always pairs with guanine). The ladder is very long; just one of our 46 chromosomes can have hundreds of millions of rungs. Under the right conditions, stretches of the ladder—sequences of nucleotide base pairs—can dictate the creation of particular amino acids that, when strung together and folded, make proteins. These proteins make it possible for an organism to go about every facet of its business. Protein-coding ladder sections are known as genes, and when geneticists isolate one of these sections they give it a unique identifier.

The plot thickens because the specific composition of any given gene can be different from one individual to another. Maybe one person has a gene in which at one locus the nucleotide sequence is ATCG, but another person has a sequence that is ATGC. (This is known as a single nucleotide polymorphism, or SNP, since only one nucleotide is different.) Or maybe one person has a sequence that reads GTCGTCGTC, but another person's reads GTCGTC. (This is known as a variable number repeat.) Either way, the result is likely to be differences from person to person in the amino acid sequence that could alter the protein that in turn could alter the behavior. These distinct versions of a gene are called alleles (an important term for the discussion to follow), and a genetic region that has different alleles is known as a polymorphic region.

One such stretch of the ladder goes by the catchy name of avpr1a, and it is getting a good deal of attention because of what has been learned from voles. Avpr1a codes for vasopressin receptors in the brain. Vasopressin (sometimes referred to as avp) is a peptide hormone that does a number of things including regulating social behaviors, particularly in males. Vasopressin receptors allow vasopressin to be "received" by key parts of the brain. Regardless of how much vasopressin is floating around the brain, without these chemical receptors, it cannot do much to regulate social behavior or anything else. Just as passengers leaving a train need a gate in order to enter the station and proceed with their business, else they wander aimlessly and unproductively on the platform, vasopressin needs receptors if it is going to activate key sections of the brain. Research within and across vole species has found a systematic relationship between the density of these vasopressin receptors and paternal behavior; voles with more vasopressin receptors are more likely to be Ward Cleavers and less likely to be Don Juans.

Genetics enters this particular story not because of variations within the avpr1a gene itself. For a gene to be expressed and actually produce the protein intended, a variety of conditions, enzymes, and other substances need to be present. If the mix of chemicals and conditions is not right, gene expression will be diminished, sometimes dramatically. In the case of avpr1a, the result would be fewer vasopressin receptors. One of the critical variables affecting expression of avpr1a pertains to a nucleotide sequence about 500 nucleotides away (upstream) from the actual avpr1a gene in what is called a "flanking region." (Avpr1a has 1,623 rungs, or nucleotide base-pairs, making it a relatively small gene.) The length of the pertinent portion of the flanking region varies in voles from around 710 nucleotides to 760. This turns out to be important, since the longer the flanking region the more vasopressin receptors are produced. Parental behavior in voles (specifically pup licking) can be fairly accurately predicted by knowing whether a vole has a long version (allele) or a short version (allele) of the genetic sequence in the flanking region. This is just one example of the way in which the effect of genetics on social behavior has been mapped.[2]

Another example is provided by fruit fly sex. We noted in Chapter 4 that fruit fly drinking habits are genetically influenced. Well, so are their mating rituals—which happen to be quite regimented. The male approaches the female, taps her on the leg, sings a song by rubbing his legs together, licks her on the abdomen, and then tries to mount her. Only males behave in this fashion—no females "in the wild" do the tap-rub-lick-mount routine. Splice the male version (allele) of a particular gene into a female, though, and she will tap, rub, lick, and mount just like one of the boys. This is an example of a single gene controlling the most evolutionarily important social behavior of any species: reproduction.[3]

Additional evidence of genes influencing behavior comes from honeybees, a very different flying insect. Bees are interesting because unlike fruit flies, they are intensely social; they live in colonies with a highly developed division of labor.[4] This division of labor means they engage in different behaviors. What explains these behavioral differences? Some of it is due to life cycle. When honeybees are first born they cannot fly or sting, so spend their days cleaning cells, grooming, or doing nothing. At about 4 days, most bees graduate to "nursing" tasks such as feeding the young and caring for the queen. At 12 days, honeybees are entering middle age and switch to construction and maintenance, sometimes transporting nectar within the hive. After mid-life, bees often leave the hive to become foragers—gathering the pollen, water, and nectar needed to produce the approximately 20 kg of honey the colony requires to survive the winter. A subset of foragers serve as scouts who go off on their own in search of either new food sources or new hive sites.

Scouts are a particular interest of Gene Robinson, Director of the Institute for Genomic Biology at the University of Illinois and one of the world's leading experts on honeybee behavior. Though bees go through a typical task progression over the course of the life cycle, they do not engage in all tasks to the same degree. Some take on particular tasks earlier and with more relish than others. This sort of variation is especially marked with regard to scouting behavior. Food scouts make up approximately 5 to 25 percent of the total foragers and hive scouts are even rarer, constituting less than 5 percent of all foragers. Many foragers never scout and some are almost exclusively scouts. Why?

Robinson and his colleagues first documented a strong overlap between hive scouts and food scouts, suggesting the existence of a scouting or, given the nature of the task, risk-taking personality in bees. Wanting to know what pushes only some toward risky scouting behavior, they took a cue from research on the molecular basis of human risk-taking. Genetic variation relevant to the dopamine reward system in humans is associated with novelty seeking and risk-taking. Honeybees do not have dopamine as such but they do possess a similar substance known as octopamine. The Robinson Lab found that, relative to other foragers, more octopamine was being produced in the brains of scouts. This finding indicates that scouts are different either with regard to their DNA, the enzyme soup necessary to initiate gene expression, or with regard to epigenetic factors that can promote or inhibit gene expression. Either way, molecular mechanisms have been found that produce important behavioral variation.[5]

Shifting from insects to mammals opens up an entirely different approach to demonstrating that genetics affects behavior: artificial selection. For thousands of years, humans have bred animals for the purpose of developing certain traits. Some of the desired traits are physical, but some of them are behavioral and temperamental. One of Charles Darwin's favorite examples of the heritability of behavior was dogs because of the huge behavioral variation across breeds. Pointers point, herders herd, and hunters hunt; these are all innate behaviors created by selecting breeding stock on the basis of ability to perform desired tasks. Darwin reasoned that if genetics was irrelevant to behavior, such selective breeding would not work—all breeds of dogs, regardless of genetic heritage, would be equally trainable on any task. You should be able to train, say, any random basset hound to herd cattle efficiently (good luck with that).

A Soviet-era geneticist called Dmitri Belyaev has provided one of the most fascinating demonstrations of the genetic basis for behaviors in animals. Belyaev quickly fell out of favor when Moscow officialdom became smitten with unsubstantiated ideas regarding the possibility that acquired traits could be inherited.[6] In 1948, Stalin and friends declared classical genetics a pseudo-science and Belyaev was fired from his job at the Central Research Laboratory in Moscow. In order to diminish the odds that further damage could be visited upon him, Belyaev packed himself off to Siberia (better to go than to be sent) where he pretended to be doing work on animal physiology when in fact he began a research agenda that significantly advanced the field of genetics.

This agenda focused on the silver fox. Because of the desirability of their pelts, large fox farms raised these beautiful animals; however, Belyaev's interests rested not with fox fur but fox demeanor. He observed that most silver foxes are quite aggressive and distraught in the presence of a human—but a few are not. He began systematically recording "flight distance," the distance between an individual fox and an approaching human when the fox would decide to "take flight." Belyaev wanted to know if there was a genetic basis for variation in flight distance from fox to fox, so he interbred those foxes with the shortest flight distance. After just 10 generations of this selective breeding, Belayev had created something quite remarkable: silver foxes that were not just tolerant but solicitous of human affection—affection the fox readily returned by wagging its tail and licking the human's hands. Interestingly, even though Belyaev only selected foxes on the basis of their behavior, physical traits, including curly tails; white coloration on the face, chest, and paws; perky ears; short jaw bones; and blue eyes also distinguished the domesticated fox that Belyaev's breeding program had created. They had so many doglike characteristics that they were soon much in demand as pets.[7]

Note that Belyaev was running a breeding and not a training program. No one conditioned the domestic foxes to behave a certain way; it was simply their innate temperament. There could hardly be a clearer indication of the relevance of genetics to behavior than the changes Belyaev induced in his silver fox population all because of selective breeding. If artificial selection can shape behavior this dramatically, natural selection can as well.

We could cite additional animal examples but the general idea is apparent. The notion that genes can influence behavior is not wild-eyed conjecture. It has been repeatedly demonstrated using different methods and in different species. Why should our species be any different? Some might argue that we are just too complex and intelligent and that any genetic influence on our social behavior is trivial compared to the influence wielded by our culture and our own free will. Or at least it is trivial when it comes to higher order behaviors like politics, religion, and morality. Well, first, as anyone who has witnessed a recent political campaign can tell you, politics is anything but higher order. Second, and somewhat paradoxically, our evolved high intelligence probably is better at predisposing us to discount genetic influences on behavior than to discern the true reasons we behave the way we do and hold the political beliefs we do. Our big brains are better equipped to spin a story that denies the relevance of biological forces than to negate biological forces. Thus, it is time to wrap our big brains around this: Human political behavior has more in common with the behavior of voles, flies, bees, dogs, and foxes than most people are ready to admit.

Politics in the Genes?

Jim Weaver represented the Fourth District of Oregon in the U.S. House of Representatives from 1975 to 1987. His grandfather, James K. Weaver, was the Populist Party nominee for president in 1892, so politics ran in Weaver's family. While in Congress, the younger Weaver earned a reputation as an irritant and a bit of a loner but also an effective legislator. A dogged proponent of environmental causes and opponent of nuclear power and the logging industry, Weaver almost single-handedly placed more than one million acres of Oregon forest land into wilderness preservation. His congressional career evaporated in 1986 when he first decided to seek the Oregon Senate seat held by Bob Packwood and then, after losing $80,000 in campaign funds speculating on commodities, was charged by the House Ethics Committee with personal use of campaign contributions. He quickly withdrew from the Senate campaign and from public office.

With his political career over, Weaver felt free to pursue what he referred to as his "obsession": the belief that politically the human species can be divided into "two very different kinds of people." In a book developing this thesis he sometimes labels these types liberal Democrats and conservative Republicans, but generally prefers the terms "ethnocentric hawks" and "empathic doves."[8]

Similar to the arguments we made in earlier chapters, Weaver sees these underlying differences in political temperament as products of "different sets of emotion" that are visible throughout human history and fiction and that are readily apparent to any "political novitiate."[9] As a politician Weaver felt so "knocked around" by these "antagonistic creatures" that he became "bound and determined to find out who they were and how they came to be."

Weaver theorizes that the core distinction between ethnocentric hawks (conservatives) and empathic doves (liberals) is attitudes toward out-groups. He describes the two groups thusly: "[O]ne is aggressive, patriotic, and insensitive to the plight of others; the second is thoughtful, compassionate, and imaginative." Weaver is a left-of-center Democrat, so no prizes for guessing which group is thoughtful and which one is insensitive. Still, Weaver's primary interest is in figuring out where these types come from rather than declaring one type better than the other. He quickly observed that some rich people were empathic doves and some were ethnocentric hawks and that the same variation existed among the economically downtrodden; thus it seemed to him that these distinct types were "not derived from economic class or personal position." What does that leave? Weaver is convinced the answer is genetics. He readily concedes that he has no idea which particular genes are relevant, but he is certain that "some mysterious genetic structures within us seem to induce tendencies toward war-like behavior in some . . . and empathic compassion in others."

Weaver's arguments are fascinating because they are made by an actual politician; his conclusion that political temperament is genetically rooted is based on long and careful observation of the hyperpoliticized world he inhabited. Unfortunately, we disagree with him on several key points. For starters, political temperaments are not divided into just two distinct camps but rather spread out along an infinite number of locations between at least two poles. The difference between two groups versus degrees along a continuum potentially has big implications for looking at the genetics of political orientation. If political temperament was genetically based and created a distinct you-either-are-or-you-aren't dichotomy, it would suggest a single gene is underlying this difference. Dichotomies result when a single gene shapes a characteristic. A person either has Huntington's disease—a known monogenetic condition—or does not have it. Characteristics that have many subtle gradations (height and IQ, for example) are the result of numerous genes interacting with each other

and with the environment. Given the wide variety of political beliefs array-
ing on multiple spectrums, few doubt that a plethora of genes, along with a
plethora of environmental factors, influence political orientations.

A second feature of Weaver's argument that seems to miss the mark is his
belief that political orientations are genetically based but do *not* run in families. Or, as he puts it, "[Y]ou may be an empathic; your brother or sister an
ethnocentric."[10] Weaver is absolutely right that siblings sometimes have very
different political orientations, but political temperaments do in fact run fairly
consistently within families and thus are not randomly distributed. Biological
siblings and parent-child dyads are more likely to share political beliefs than
any two randomly selected individuals. This political resemblance of genetic
relatives could be due to pure socialization—that is, to mom and dad conditioning the kids to be good conservatives or liberals—but also could be due to
genetic influence.[11] After all, the kids have mom and dad's genes and are more
likely than not also to have their political beliefs. So, while a positive correlation
exists for political beliefs within families, the source of that correlation—genes,
family environment, or some combination of those two—is uncertain.[12]

Is it possible to determine whether political orientations have a genetic component? Yes, but assessing the role of genetics in humans is more challenging
than it is for foxes, dogs, honeybees, and voles. Legally, ethically, and morally
we cannot perform the same kinds of tests on humans that are conducted on
fruit flies and silver foxes. Gene splicing and selective breeding for the edifi-
cation and amusement of geneticists and social scientists obviously is taking
intellectual curiosity too far. Researchers on genetics and human behavior are
largely restricted to naturally occurring, not manipulated, situations. Rather
than creating liberals or conservatives by controlling mate choice or splicing
a certain genetic allele into a person's DNA, we must be content with trying
to figure out whether liberals and conservatives have different alleles—or if
people who are more similar genetically also tend to be more similar politically.

Bring on the Clones

Like Kay Rene Reed and DeeAnn Angell, Jim Springer and Jim Lewis were born
on the same day in the same place, met each other as adults after a shocking
family revelation, and had DNA that suddenly allowed previously mysterious

situations to make sense. Also like Kay Rene and DeeAnn, Jim and Jim discovered they had a good deal in common, only in their case not with each
other's families, but with each other. They looked alike. They had similar body
language—crossing their legs the same way, for example. They had similar personalities (patient, kind, and serious). In school they had both loved math and
loathed spelling. They both had the same favorite vacation spot (Pas Grill Beach
in Florida). They both enjoyed carpentry as a hobby and had occupational his-
tories in law enforcement. They both gained 10 pounds at exactly the same age
but had otherwise stable weights. Beginning in their teens and ever since, they
both suffered from migraines. They both experienced serious heart problems in
their thirties. They had nonverbal IQ scores just one point apart (though their
overall IQ differed a little more). A team of psychologists gave them a battery
of 23 vocational tests and discovered they were so similar it was as if the same
person had taken each test twice. It was almost like they were twins.[13] In fact,
they were. They had been adopted at the age of one month by different families
who were both told their new child was a twin but that the sibling had died.
Some years later, when Jim Lewis's mom was filling out legal paperwork related
to the adoption, a court officer blurted out that her son's brother was alive and
well. She relayed this fact to her adopted son but he was not eager to act on this
information and she respected his decision. It was not until he was 39 that Jim
Lewis changed his mind and the brothers actually found each other and began
discovering similarities. In some respects, these similarities are not so surprising because the Jims are not just twins but a special type of twin.

Sex education time. Conception occurs when a sperm enters an egg, the
nucleic DNA of the sperm and egg merge, and a single zygote is formed. This
single-celled zygote soon divides and begins to grow into what will become a
multicellular human being. That is the usual course of events. Occasionally,
somewhere between several hours and a couple of weeks after conception, for
reasons not fully understood, the single zygote splits into two and both zygotes
eventually grow into full-fledged human beings. Because the resulting twins
started as the same zygote, the genetic heritage of these siblings is virtually
identical. Routine and ongoing mitotic cell division will certainly create some
genetic variation between them over the years, but twins that form from a single zygote (called monozygotic or MZ twins) are basically genetic clones. This
makes MZ twins like the Jims unique in a number of ways. To take just one

example, MZ twins are as closely related to their nieces and nephews as they are to their own children. More important for our purposes is the fact that MZ twins can be used as a baseline for assessing the role of genetics in a wide variety of traits. This kind of research is known as a twin study.

The basic idea of a twin study is to compare MZ twins with a completely different type of twin pair—and such a type is indeed available. Unlike MZs, dizygotic (DZ) twins are a product of two separate eggs being fertilized by different sperm. The result is two zygotes growing in the same womb at the same time, but these zygotes were always separate. Genetically, DZ twins are no more similar than any other pair of full siblings (on average, this similarity is 50 percent). Like most MZ twins, though not Jim and Jim, DZ twins tend to be raised in the same environment—same mom, same dad, and same socioeconomic class. The analytical value of the two types of twins is that they make it possible to get leverage on the extent of genetic influence. If a trait, such as height, is measured and it is found that pairs of MZ twins are more alike than pairs of DZ twins, a genetic influence is suggested. This is because the big environmental forces that push MZ twins to be more like each other are the same forces that push DZ twins to be more alike: that is, the environment they share due to the fact that they are siblings in the same family. The underlying logic is that if height is purely a product of environmental influences—nutrition, exercise regimen, and so forth—the correlation between MZs and DZs would be roughly the same. On the other hand, if MZ twins are more alike than DZs, genetics probably is one of the reasons.

By comparing the similarity of MZ twins with the similarity of DZ twins on a given trait, twin studies indicate the extent to which that trait is genetically influenced. Twin studies, however, say nothing about the particular genes that are doing the influencing. Twin studies have been used in the medical field for decades to identify the particular illnesses and conditions that have a heritable component and, therefore, that might benefit from further research at the genetic level. For example, some of the early work on the genetics of breast cancer involved twins and served as a justification for pursuing molecular efforts to identify the genes involved. These sorts of contributions to health research are one reason most states and many countries have twin registries. Twins in these registries are constantly being pestered with surveys and pleas to participate in experiments to help advance the cause of science.

Though early twin studies dealt primarily with health-related issues, social scientists increasingly use them to test whether or not behaviors and attitudes have a heritable basis. It was in this vein that more than a quarter of a century ago Nick Martin and Lindon Eaves published a landmark article.[14] Martin and Eaves placed the Wilson Patterson index—the measure of conservatism we have referenced in previous chapters—in a survey of thousands of twin pairs and found that the social and political attitudes of MZ twin pairs were indeed more similar than those of DZ twin pairs. The blockbuster implication, of course, is that social attitudes, including political temperament, are genetically influenced. Prior to Martin and Eaves there had not been much interest in asking twins about their attitudes because it seemed "obvious" that social attitudes generally, let alone political attitudes specifically, could not be heritable.[15] The Martin-Eaves finding suggested a way of thinking about politics and political attitudes that was quite foreign to those who had been trained to assume that environmental factors are the basis of all political beliefs.

In the past 10 years or so, the discipline of political science has begun to catch up with the findings flowing from behavioral genetics and there has been something of a boom in research applying twin studies to political attitudes.[16] The basic gist of these studies is summed up by Figure 7.1. This shows two scatterplots, one for MZs and one for DZs. In each graph, the Wilson Patterson scores representing overall position on a collection of political issues for one twin are plotted against the other twin. These data are taken from a 2009 survey of a twin sample we conducted in conjunction with researchers at the University of Minnesota, and the scores themselves are scaled so that positive numbers reflect more conservative issue positions and negative numbers reflect more liberal issue positions. Note that the relationship for MZs is immediately discernible; indeed, the correlation for the MZ data is 0.62. The relationship for DZs is there, but it's much looser in comparison. To be specific, the correlation for DZs is 0.35, or roughly half of what it is for MZs. We found this basic pattern held for all sorts of measures of political temperament measures, including self-reported ideology, right-wing authoritarianism, and the Society Works Best index, in addition to the Wilson Patterson index of specific issue positions. Political temperament, in short, seems to be at least partially heritable.

Other studies conducted on different twin registries around the world uniformly report similar findings. Precise estimates of the degree of genetic

MZ Twins

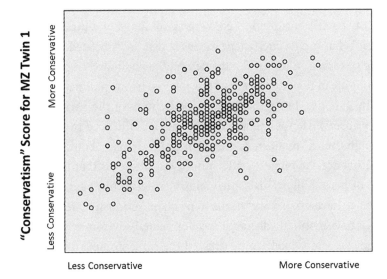

"Conservatism" Score for MZ Twin 2

DZ Twins

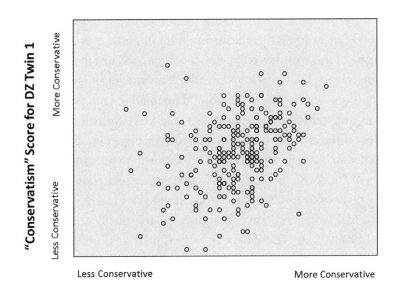

"Conservatism" Score for DZ Twin 2

Figure 7.1   Similarity of Issue Positions for MZ and DZ Twins

influence on political temperament reported by twin studies should be taken with a grain of salt since they vary from sample to sample. Generally, though, heritability coefficients tend to converge around 0.4 or a little less for political attitudes. What such a coefficient means is that an estimated 40 percent of the variance observed in political attitudes can be attributed to genetic influence. It does not mean that 40 percent of any given person's political attitudes is controlled by genetics, but rather that a good chunk of the variation in political attitudes observed across people is likely being influenced by genes.

Though much replicated, these findings remain controversial. Within political science the biggest criticism of twin studies is that they overestimate the role of heritability. This same general criticism has been made as long as twin studies have existed and needs to be taken seriously. The argument is that MZ twins have distinctly different developmental environments than DZ twins since they are more likely to be dressed the same, to take the same classes in school, to have the same friends, to share the same bedroom, to interact with each other more, and to be mistaken for each other. If the environments of MZ twin pairs are more similar than the environments of comparable DZ twin pairs, estimates of heritability may be picking up the more similar environment of MZ twins in addition to their more similar genetics.

Much ink has been spilled on such potential environmental confounds and the general conclusion is that the assumptions underlying twin models are safe. We did one of these studies ourselves. Our approach was to assume that, if the critics are correct, DZ twins who are quite similar when it comes to having the same bedroom and the same friends and taking the same classes should be as similar on the trait of interest as the average MZ twin pair. Similarly, we reasoned that if the critics are correct, MZ twins who are very different on these environmental characteristics should look more like the average DZ twin pair in terms of similarity. Instead, the results showed that variations in environmental similarity do not make much of a difference, at least in terms of political attitudes. MZ twins with very dissimilar environments are still much more alike politically than DZ twins with very similar environments. The genetic influence stayed stubbornly consistent even when we replicated our analysis on two different twin data sets. Apparently, being dressed the same, sleeping in the same room, having the same friends, and taking the same classes has little independent bearing on the eventual similarity of adult political views. The key inference is that the twin-generated heritability estimates for political attitudes are not unduly biased by the greater environmental similarity of MZ twins.[17]

But don't take our word for it. Remember Jim Springer and Jim Lewis, the identical twins adopted by different families as newborns? It is difficult to find a single set of twins like that, let alone a bunch of them. If by hook or crook you can put together a data set of twins raised apart, though, you have a terrific basis for doing heritability research. The basic logic of a twins raised apart study is that if people like the Jims (MZ twins reared apart) are similar on a trait, that similarity is much more likely to come from what they have in common than what they do not have in common. What the Jims obviously share is their genetic heritage and not their family environments (after all, they were raised in different homes). This is a powerful research design for studying genetic influence since it does not require any assumptions about equality of environmental influence. Unfortunately, because of understandable difficulties in data collection, studies of twins raised apart are relatively rare, and studies that examine the political traits of twins reared apart are like hens' teeth. If such studies produced political temperament heritability estimates that are similar to the recent raft of twin studies in political science, however, it would be strong confirmatory evidence of a genetic influence on political temperament.

As far as we are aware, only one study has actually done this. The Minnesota Study of Twins Raised Apart ran for 20 years and included extensive testing on 81 MZ twin pairs and 56 DZ twin pairs. Thomas Bouchard, one of the principals behind this work, asked these twins questions about their issue positions (a version of the Wilson Patterson) and their attitudes toward authoritarianism (Altemeyer's right-wing authoritarianism items). Estimates of heritability on these measures came very close to those generated by conventional twin studies. If anything, the heritability estimates were a bit higher; for example, the heritability of the Wilson Patterson battery of general conservatism was calculated at 0.60.[18]

Another research design that can get at genetic influence without needing to know anything about genetics at the molecular level draws on data from adoptions. The basic reasoning is that if adoptees have similarities to their biological parents in addition to their adoptive parents, then biology must be involved. Adoption studies are difficult to conduct because they require access to the adoptee, the adoptive family, and the biological family, which makes data collection a challenge. We do not know of any adoption study that looks specifically at political attitudes, but a couple analyze participation in politics (like political temperament, twin studies have also found political participation to be genetically influenced) and report a solid correlation between adopted-away offspring and their biological parents.[19]

The central finding of research in this area, regardless of design, is that genetics plays a role in shaping not just physical appearance and not just *aptitudes* for music, sport, or sums, but also in shaping *attitudes* toward music, sport, and sums, not to mention toward the bedrock dilemmas of politics. Please note that none of these studies suggests that genes by themselves *determine* ideology, as they all leave plenty of room for socialization, culture, and unique life experiences to exert important influences. If research on political temperament follows the path of health research, we would now be at the point where bright enterprising researchers take advantage of the tools of molecular genetics to begin hunting for the genes that shape political views. And so it has come to pass, though efforts to find these are just beginning and to this point only a few published works are available for summary.

A Liberal Gene? It Doesn't Work That Way

Two major approaches have been used to investigate the links between genes and traits. One involves scanning hundreds of thousands of sites where genetic variations (polymorphic regions) are known to exist. These scans do not cover the whole genome (that is, every single nucleotide), but they do include sites scattered across the genome, and therefore are known as "genome-wide." The question becomes whether variation at any of these sites correlates with the trait of interest. Genome-wide association studies (GWAS) do not identify specific genes that vary with the trait but they can identify regions of the genome where those genes are likely to be found, a useful advance because it reduces the size of the haystack by creating a more manageable number of genes to inspect.

Pete Hatemi is a political scientist who specializes in the genetic influences on political temperament, and he spearheaded the first genome-wide study specifically aimed at political attitudes and behavior. Hatemi and his colleagues found four chromosomal regions that seem to correlate with ideology. Those regions are known to include a number of genes related to the regulation of social behavior, which makes sense given that politics is a form of social behavior.[20] Interestingly, several of the regions identified in Hatemi's study are linked with olfaction (the process of smelling). This might seem a little odd, but recall from earlier chapters the preliminary indication that sensitivity to certain smells (think disgust) and preferences for certain foods (olfaction is related to taste) correlate with variations in political beliefs.

The other approach is to test whether genetic variation at a single, pre-identified site correlates with political temperaments. Only a handful of these candidate gene association studies examine variations in political temperament, and only one attempts to identify the genetic loci of ideology. It was conducted in James Fowler's lab at UC–San Diego, with Jaime Settle as the lead author (along with Hatemi, Fowler is one of the acknowledged leaders of the nascent genopolitics movement).[21] The candidate gene singled out in this case is called DRD4 since it is known to relate to the D4 class of dopamine receptors, key parts of the so-called dopamine reward system. Parallel to the vasopressin receptors relevant to vole behavior, dopamine receptors have been shown to be relevant to human behavior. More specifically, an allele of DRD4 in which a particular nucleotide sequence repeats seven times (known as the "long" allele) has been connected in many earlier studies to exploratory behavior, sensation seeking, and even ADHD (notice the parallel to the research on the particular molecular features of the reward system in risk-taking honeybee scouts).

Settle and colleagues reasoned that, since previous research repeatedly finds that liberals are more likely than conservatives to seek new experiences (see Chapter 4), they may also be more likely to have the long version of DRD4. Their initial test did not reveal the expected relationship, but further analysis suggested that the pattern did materialize among people who reported having many friends when they were children, thereby raising the possibility of a gene-environment interaction in which the impact of the gene on the trait of interest depends on environmental conditions. Settle et al. argue that this makes sense since they believe friends are necessary for a propensity for doing things (provided by the long version of DRD4) to be translated into actual new experiences and ultimately into political liberalism.[22]

Though the research connecting political temperament to genetics at the molecular level is suggestive, those involved in it would be the first to tell you the results need to be viewed as tentative. The impact of any single gene on something as complex as political attitudes and behavior is likely to be pretty small, and even then that impact is likely to be dependent upon interactions with other genes (as is the case with DRD4) or with environmental conditions. Moreover, avoiding false positives is a special problem for GWAS since they check for so many possible correlations that a few are going to look promising even though they occurred merely by chance. Perhaps this is part of the reason that findings from molecular genetic studies have not replicated well regardless of whether the trait of interest is drug addiction, breast cancer, or personality.[23]

The researchers publishing in this area are careful to attach the appropriate caveats to their findings, but the media and blogosphere have an annoying habit of ignoring caveats. The Settle et al. study, for example, generated headlines trumpeting discovery of a "liberal gene" even though DRD4 is better described as a gene, like many others, that is relevant to a certain class of dopamine receptors that in turn is relevant to all kinds of behaviors and conditions. For example, the dopamine reward system is a key factor in controlling muscle movement (and therefore is a factor in Parkinson's disease) and also in directing organisms toward pleasurable situations (and therefore is relevant to addictive behaviors). The study actually found that allelic variations in DRD4, when interacting with certain environmental factors, might (if the findings are replicated in future studies), along with many other variables, be relevant to political views. That is not nearly as sexy as "Scientists Discover Liberal Gene" but it is much more reflective of the research.

An additional illustration of the value of molecular genetics is provided by the work of geneticist Peter Visscher and colleagues. Visscher developed a way of estimating heritability that does not require twins, adoptees, or questionable assumptions about environmental similarity. It works like this. Full sibling pairs who are not MZ twins share 50 percent of their genetic heritage on average. The optimal phrase in the previous sentence is "on average," however, and some pairs have been found to share as little as 38 percent, and others as much as 62 percent (confirming casual observation that some siblings are really alike and others less so). Similarly, variable degrees of genetic concordance (though varying around a much lower mean) are detectable in any collection of pairs of randomly selected people. The Visscher technique uses genome-wide scans to quantify the degree of genetic similarity of a pair of individuals and reasons that if a trait of interest is more similar among pairs that are genetically similar than among pairs that are genetically dissimilar, the trait must have some basis in genetics. Visscher tested the technique on a trait known to be heritable—height—and obtained a heritability coefficient that very nearly matched that produced by twin studies (just under 0.80).[24] Interestingly, when the technique was applied to political attitudes, the heritability coefficient was approximately 0.2, less than the roughly 0.3 to 0.4 commonly reported in twin studies. Still, note that the estimate is significantly greater than zero and that the procedures employed deprive critics of their standard complaint that the assumptions behind twin studies are questionable (remember, twins are not employed in the Visscher procedures).[25] Of course, 0.2 is a far cry from 1.0, meaning that political beliefs also must emanate from sources other than common single nucleotide polymorphisms. What might these other factors be?

Environment, Environment, Environment

London is one of the largest and most complex metropolitan areas in the world, covering 620 square miles of southeast England and called home by more than 8 million people, just over 12 percent of Britain's total population. When 2012 Republican presidential nominee Mitt Romney expressed reservations that London was fully ready for the 2012 Olympics, British Prime Minister David Cameron was quick to remind the organizer of the 2002 Salt Lake City (population 180,000) Olympics of London's scope. "We are holding an Olympic Games in one of the busiest, most active, bustling cities anywhere in the world. Of course it is easier if you hold an Olympic Games in the middle of nowhere." Cameron, it would seem, was not overly interested in courting the Mormon vote. London's mélange of weaving, name-changing streets is legendary. No simplifying, geographically arranged, grid-like patterns here, but an evolving collection of thoroughfares, byways, alleys, drives, and courts. No one dares attempt to estimate the total number of streets, but among their number is "Bullocks Terrace," "Bird-in-the-bush Road," and "Shoulder of Mutton Alley." There are so many "High Streets" that two thirds of all Londoners live within a 5-minute walk of one.

Driving a taxi in this tangled, congested, sprawling web requires memory, spatial skills, and directionality beyond the ordinary. Neuroscientist Eleanor McGuire and colleagues wondered whether the mental abilities of London taxi drivers were reflected in their anatomical structures, so they put 16 taxi drivers, one at a time, in a brain scanner. Compared to a larger control group of sociodemographically but not occupationally comparable Londoners, the researchers found that the posterior hippocampus volume of the taxi drivers was significantly greater, a sensible result given the role of the posterior hippocampus in spatial representation and directionality. But which came first, a brain well equipped to recall and navigate a daunting nest of backstreets and suburbs, or a situation demanding that brains develop in such a way as to facilitate certain tasks? Did the mental capacity find the right job or did the job shape the mental capacity? This is a difficult question to answer definitively—the acid test requires neurological data on the same individuals over time—but McGuire and colleagues did the best they could and compared the hippocampuses of long-time and relatively new cabbies. They discovered that, even when controlling for age, the longer an individual had been a cabbie the bigger the posterior hippocampus, suggesting that environmental situations and events have a clear, measurable effect on biological constitution.[26]

It may seem odd to pick this particular place in our book to summarize the McGuire findings illustrating the power of the environment, but a true understanding of the role of genetics requires coming to terms with environmental factors. Moreover, it is time we reminded readers that biological characteristics do not need to be genetic. Within very real limits, nervous systems are plastic and physical characteristics such as the shape of individual neurons and the contours of the hippocampus can be altered as a result of the environment in which they operate. The genetic material we inherit from our parents is crucial—but, as indicated by the apparent changing nature of London cab drivers' hippocampuses, far from determinative. The DNA of the cabbies did not change after they had driven taxis for many years, but their neural anatomy seemed to. These changes are obviously biological (what is more biological than brain contours and features?) but they are the result of experience, not genes.

Those of you who need additional doses of "the environment matters too," fear not. Two of the animal stories presented earlier in this chapter have addenda that illustrate the power of environmental factors. Remember the female fruit flies that, via splicing of a single "male" allele, could be made to go through the entire male precopulatory ritual? We mentioned then that no female in the wild ever behaved in such a fashion—but guess what? "Wild type" (not genetically spliced) females in the vicinity of a female that has received the "male" version of the gene will often, after watching the altered female engage in male sexual behaviors, begin to do the same. If fruit flies could talk, it is likely we would hear the following conversation:

Maude:	What on earth has gotten into Betty?
Flo [these are typical female fruit fly names]:	I have no idea; she's never acted like this before. It is embarrassing to the sisterhood.
Maude:	Still, it does look like she's having fun.
Flo:	Well, yes it does. Do you think we should give it a go ourselves?

In this case, the gene-spliced female became part of the environment that then affected the behavior of the nonspliced female fruit flies.[27]

Now think back to the vole story. It seemed to be about genes and flanking regions, but it also illustrates the importance of nongenetic biology and development. The number of vasopressin receptors is not determined solely by the length of the avpr1a flanking region; it can also be influenced by social environment. For example, if a doting father vole is placed with a litter sired by an uncaring vole, the "guest" father will typically respond by licking the pups. After all, Ward Cleaver is Ward Cleaver even if he's dealing with Eddie Haskell rather than Wally or the Beaver. This licking, if it occurs at the appropriate developmental stage, can increase the number of vasopressin receptors in the pups quite apart from any genetic tendencies traceable to avpr1a. As a result, when they mature, these offspring are more likely to become attentive fathers themselves and their subsequent licking of pups will lead to a high number of vasopressin receptors in those offspring, and so on. Well-timed licking not only leads to behaviorally relevant biological changes but the effect is intergenerational. An entire line of attentive fathers has been molded, without any genetic change, merely because one litter of pups was licked at a crucial developmental stage.

Note that this intergenerationally transmitted behavior is not learned. If it was merely learned behavior, being licked at a later stage of development would have just as strong an effect—probably even stronger—as being licked at an earlier stage, and there would be no change in vasopressin receptor density.

Instead, this is a crystal clear example of behaviorally relevant biological differences that are not necessarily genetic. Doting vole fathers are physically different from nondoting fathers in that they have more vasopressin receptors; the cause may be genetic or the cause may be the developmental environment, but the change is undeniably physical.

Predispositions of the sort we have been discussing throughout this book are not always genetic, but they are instantiated in deep psychological and physiological patterns. These biological signatures can change but not quickly. When a long-term London cabbie decides to hang up his driving gloves in order to better enjoy the comforts of his flat, hippocampus volume does not immediately shrink, but it probably does begin to change gradually. Behaviorally relevant biological predispositions (BRBPs) are not completely innate. The nature-nurture debate is misleading and, in many respects, beside the point, a fact that is pointed out in most every popular science book published but, judging from media headlines and blog posts, does not always register. Both genetics and the environment matter and it is not necessary, or even possible, to declare one the winner. Thus, with apologies to Christine Lavin, whose creative song lyrics appear at the beginning of this chapter, "hardwired" is not an accurate way to describe the connection of genetics to behavior.

Genes Shape Looks *and* Likes

How could a behavior, whether it is Kay Rene Reed chewing her fingernails, a border collie herding sheep, a fruit fly rubbing its legs together, or Jim Weaver being empathic toward out-groups, be encoded in genetics? The answer is the same way physical traits are. Whether the trait of interest is behavior or appearance, the mechanics of genetics are the same and primarily involve chemical substances and their receptors. For example, the process leading to physical growth in an infant begins in the brain's hypothalamus, which regulates the secretion of a growth hormone in the pituitary gland (also located in the brain). This growth hormone then enters the bloodstream and flows to the liver, where it stimulates something called insulin-like growth factor-1, which causes various tissues to expand. Genetics is relevant at every stage of this process since it creates proteins. These proteins might be the growth factor itself, the chemical receptors that allow the necessary substances into the pituitary gland, or the transcription factors that help to start up the genes that lead to the production of hormones and receptors. If your version of one of these genes hinders the production, transportation, or uptake of the growth hormone, you will likely be shorter than you otherwise would have been.

The genetics of behavior is nearly identical; it is just that the particular protein and receptor locations are different. For example, the dopamine system affects behavior just as the growth hormone system affects physical height. Allelic variation in one gene (the aforementioned DRD4) is known to affect a certain class of dopamine (instead of growth hormone) receptors and, because of its influence on receptor density and efficiency, to affect the movement of the neurotransmitters that alter behaviors associated with risk-taking and ADHD.[28] Just as genetically produced variation in the nature of receptors on the pituitary gland affect height, genetically produced variation in the nature of dopamine receptors on the striatum (a part of the brain's dopamine reward system) affect behavior. If more proof is needed of the relevance of these genetically influenced neurotransmitters and receptors to behavior, consider this: Cocaine operates by activating dopamine receptors, particularly in the nucleus accumbens (yet another part of the brain's dopamine reward system). Hallucinogens like LSD and mescaline activate serotonin (and other) receptors. Does anyone deny that these chemically induced alterations in neurotransmitter systems affect behavior? Does anyone deny that genetically produced variations in these same neurotransmitter systems also affect behavior?

Genetics uses the same types of levers to shape behavior as it does to shape physical traits. As a result, it requires intellectual gymnastics of the first order to claim that genetics affects our bodies but not our brains, our appearance but not our actions (and attitudes). If genetics can influence neurotransmitters near the pituitary, it can influence them near the amygdala as well; the former affects our size and the latter our emotionality. Behaviors result from a physical process initiated by the environment and shaped by genes. Unless you believe in some mystical, ethereal, unscientific force simply because that is what you want to believe, there is nothing else that could lead to behavior.

The fact that humans can, on occasion, use other parts of their brains to mitigate the inclinations emanating from, say, their amygdalas does not alter the fact that some people will have a more responsive amygdala than others and that over the long run and with enough observations, behaviors on average are likely to be correspondingly different from one amygdalic group to the other. Kay Rene can tape over her fingernails or will herself not to chew them for a month, but this does not change the fact that genetically shaped behavioral predispositions exist. It just means these predispositions do not determine behavior. The importance of predispositions stems from the fact that most of the time we let them purr along unperturbed. This is our default system. Key parts of the brain, such as the prefrontal cortex, only swing into action when other parts of the brain, notably the anterior cingulate cortex, sound the alarm indicating that the default system may not lead to the best course of action in a particular situation. Conscious awareness is often not involved when we are on autopilot, and when we *are* conscious that we have made a decision, we are typically unwilling to admit that predispositions acted as a thumb resting surreptitiously on our decision-making scale. As Shankar Vedantam points outs, "no one intuits the presence of a neurotransmitter."[29] Perhaps this is why humans overstate the extent to which behavior is the product of willed, conscious thought.[30]

Conclusion: Darwin, Right; Wallace, Wrong

The progression of conventional wisdom is revealing. While Darwin was convinced that behavior was genetically based, his contemporary and coformulator of the theory of natural selection, Alfred Russell Wallace, was not. Wallace believed that physical traits such as the sharp teeth of beavers could be inherited but that the knowledge of what to do with those teeth could not. Darwin's view was that genetics could bestow robust teeth as well as innate predispositions to use them to fell trees and construct dams. No doubt Wallace's take on this matter came much closer to the views of most of those living in the Victorian era. Today, however, scientifically literate people accept, perhaps grudgingly, that genetics is relevant to basic behaviors such as those connected to survival, reproduction, and instincts, but many still resist a connection of genetics to the so-called higher order domains of religion, politics, and morality. Genetics was previously thought to be relevant only from the neck down, but now is increasingly recognized as also being relevant to the limbic or emotional parts of the brain.

Emotions, maybe; dispassionate decision-making, certainly not. Genetic influences on the precious frontal lobe, site of executive decision-making, are still widely resisted—but how long can this last? The vision (hope?) that the human prefrontal cortex plays by different rules than all other aspects of life on earth is increasingly untenable. After all, the prefrontal cortex is built and integrated in the same way as other parts of the brain. It is composed of the same kinds of neurons and support cells as "lower" parts of the brain. As neuroscientist David Eagleman notes, no part of the brain has been found "that is not itself driven by other parts."[31] It is only a matter of time before the prefrontal cortex also falls to biology. Darwin's advance and Wallace's retreat on the issue of the influence of genes are leaving environmental determinists with an ever-shrinking domain. Their last stand is taking place on the turf where higher-order decision-making occurs. They want very much to believe that some part of the brain is immune from basic biological principles, but neuroscientists give them no cause for glee. Behaviorally relevant biological dispositions exist; they have been constructed in part by genetics, and they permeate every part of our brain, not just those parts dominated entirely by the emotions. It is likely that Wallace's last adherents will be decamping from the behavioral sciences (including political science) within the next half century.

By way of conclusion, we return to DeeAnn Angell and Kay Rene Reed. We described them in some detail in the opening section of this chapter but never mentioned their politics. One of our students (Jayme Neiman) became deeply interested in the Angell-Reed case and sent several questions to the two "swisters" about their own political views and those of the two families. Only Kay Rene replied, but her answers are fascinating. She reported that the Reeds, with whom she was reared, thought of themselves as moderate conservatives but were actually quite conservative, particularly on certain issues. The father who raised Kay Rene, for example, was strongly opposed to women working outside the home, going so far as to "beg" Kay Rene not to take a job until her kids were grown. Kay Rene classifies herself as a political moderate but admits to clear liberal tendencies on some issues, particularly gender equality.

Are Kay Rene's occasional liberal inclinations traceable to the political views of the Angells, her biological family? Maybe, but Kay Rene thinks probably not. She attributes her views instead to the experience of going to college in the 1970s when women's rights and racial equality were highly salient issues. Moreover, though understandably less certain of the political views of the Angell family, she suspects they also tended toward the conservative side of the ledger though not nearly as much as the Reeds. Kay Rene's more liberal politics relative to her "adoptive" family could indeed be attributable to her college experiences—nothing in this chapter would necessarily contradict that. Then again, as we have seen, people's brains often build narratives that downplay the role of behaviorally relevant predispositions that are in part genetic. If you had been raised in an entirely different environment, would your political orientations be completely unrecognizable from what they currently are? If Al Franken and Rush Limbaugh had been switched at birth, would Franken be conservative and Limbaugh liberal? Or, resting at the core of our beings, are there features of our personality, physiology, and politics that would peek through no matter the environmental context in which we find ourselves? Burgeoning empirical research suggests we should not dismiss the latter possibility too readily.

Notes

[1] Barville, "Pair Were Switched at Birth"; and Ibanga, "Switched at Birth: Women Learn the Truth 56 Years Later."

[2] Hammock and Young, "Microsatellite Instability Generates Diversity in Brain and Sociobehavioral Traits."

[3] Demir and Dickson, "Fruitless Splicing Specifies Male Courtship Behavior in Drosophila."

[4] Johnson, "Division of Labor in Honeybees."

[5] Liang et al., "Molecular Determinants of Scouting Behavior in Honey Bees."

[6] The leading player in this movement was an agronomist named Trophim Lysenko.

[7] Trut, "Early Canid Domestication: The Farm-Fox Experiment"; and Goldman, "Man's New Best Friend."

[8] Weaver, *Two Kinds: The Genetic Origin of Conservatives and Liberals.*

[9] They also parallel Haidt's argument that political orientations are at least in part based in "intuitive ethics," reflexive judgments of right/wrong. See Haidt, "The Righteous Mind."

[10] David Lykken suggests the trait of genius is genetic but does not run in families, though his logic is different from Weaver's. Lykken suggests that genius is a configural trait, requiring just the right combination of genetic alleles. Many of these alleles, he reasons, are present in intelligent people, but real genius demands a very specific combination of these alleles. Thus, Einstein's ancestors and progeny are likely to be intelligent but not geniuses (Lykken, "The Genetics of Genius").

[11] Jennings and Niemi, "The Transmission of Political Values"; and Niemi and Jennings, "Issues and Inheritance in the Formation of Ideology."

[12] For further discussion of politics and genetics, see Jimenez, *Red Genes, Blue Genes: Exposing Political Irrationality*; and Haston, *So You Married a Conservative: A Stone Age Explanation of Our Differences, a New Path Towards Progress.*

[13] Segal, *Born Together—Reared Apart: The Landmark Minnesota Twin Study.*

[14] Martin et al., "The Transmission of Social Attitudes."

[15] Bouchard and McGue, "Genetic and Environmental Influences on Human Psychological Differences."

[16]   See Ibid., "Genetic and Environmental Influences on Human Psychological Differences"; Alford et al., "Are Political Orientations Genetically Transmitted?"; Fowler et al., "Genetic Variation in Political Participation"; Hatemi et al., "The Genetics of Voting: An Australian Twin Study"; Klemmensen et al., "The Genetics of Political Participation, Civic Duty and Political Efficacy across Cultures: Denmark and the United States"; and Bell et al., "The Origins of Political Attitudes and Behaviours: An Analysis Using Twins."

[17]   Smith et al., "Biology, Ideology, and Epistemology: How Do We Know Political Attitudes Are Inherited and Why Should We Care?"

[18]   Bouchard et al., "Genetic Influence on Social Attitudes: Another Challenge to Psychology from Behavior Genetics."

[19]   Cesarini et al., "Pre-Birth Factors and Voting: Evidence from Swedish Adoption Data."

[20]   Hatemi et al., "A Genome-Wide Analysis of Liberal and Conservative Political Attitudes."

[21]   Settle et al., "Friendships Moderate an Association between a Dopamine Gene Variant and Ideology."

[22]   Another intriguing line of research investigates the connection of various genetic loci with political participation rather than liberal-conservative ideology. In "Two Genes Predict Voter Turnout," Fowler and Dawes find that genes related to serotonin and MAOA may be related to variations in voter turnout, though subsequent research raises questions about the robustness of this relationship (see Charney and English, "Candidate Genes and Political Behavior"; and Deppe et al., "Candidate Genes and Voter Turnout").

[23]   Goldstein, "Common Genetic Variation and Human Traits."

[24]   Visscher et al., "Assumption-Free Estimation of Heritability from Genome-Wide Identity-by-Descent Sharing between Full Siblings."

[25]   Benjamin et al., "The Genetic Architecture of Economic and Political Preferences."

[26]   Maguire et al., "London Taxi Drivers and Bus Drivers: A Structural MRI and Neuropsychological Analysis."

[27]   Demir and Dickson, "Fruitless Splicing Specifies Male Courtship Behavior in Drosophila."

[28]   Kunen and Chiao, "Genetic Determinants of Financial Risk Taking"; and Dreber et al., "The 7R Polymorphism in the Dopamine Receptor D4 Gene Is Associated with Financial Risk Taking in Men."

[29]   Vedantam, *The Hidden Brain*, 44.

[30]   For example, humans frequently engage in magical thinking, meaning they believe they can control things over which they have no control (the number that will come up on dice they are about to roll). Also, when the motor cortex is stimulated in such a way as to cause a person to stand up, the person will often claim to have stood up because of a desire to stretch or see better, when in fact this was not the reason at all. We have an ingrained desire to spin a rational narrative to account for our actions, but major parts of this narrative are a fiction.

[31]   Eagleman, *Incognito*, 166.

CHAPTER 8

The Origin of Subspecies

*You can't Belong among us unless you Believe what we Believe . . .
[and] . . . if you don't Belong among us, then you are our inferior, or
our enemy, or both.*

Tom Robbins

*The conclusion that all humans are effectively the same is
unwarranted . . . evolution has taken a different course in different
populations.*

Gregory Cochran and Henry Harpending

Do you agree or disagree with the following statement? "Human beings,
as we know them, developed from earlier species of animals." Actual
responses taken from polls conducted in 34 countries between 2001 and
2005 make for a fascinating comparison of attitudes toward evolution.[1]
There is little controversy in the most developed countries included in
the survey. For example, in Iceland, Denmark, Sweden, France, Japan, the
United Kingdom, and Norway, between 80 and 90 percent of the population agrees that "humans developed from earlier species." On the other
hand, in countries with lower levels of development and education agreement sometimes dips below 50 percent. The five countries on the low end
of the "support evolution" spectrum include Latvia, Lithuania, Cyprus, and

Turkey, with Turks being the most skeptical: Only 23 percent of them agree with the statement.

What is the country rounding out the bottom five, a country that ranked just above Turkey in its skepticism of evolution? Astonishingly, it is arguably the most educated and economically developed country on the planet. It is acknowledged as the world leader in scientific accomplishments, it spear-headed the development of nuclear power, is the only country to put people (12 of them) on a celestial body, and boasts 338 Nobel laureates (nearly three times as many as the country with the next most—the United Kingdom), and year after year attracts undergraduate and graduate students from all over the world to study biology and medicine at its world-leading universities. Yet only 40 per-cent of the citizens of the United States believe humans developed from earlier animals. Significant portions of the remaining 60 percent are convinced that humans burst on the scene in their current form, shape, and size approximately 6,000 years ago and have not changed since. Any way you slice the numbers, a good chunk of Americans simply do not believe the most basic and rudimen-tary tenet of modern biological science: the evolution of species.

This situation might be slightly more excusable if Americans actually knew what they were rejecting. Yet the denial of evolution is accompanied by a remarkable level of ignorance concerning evolution's basic principles.[2] Maybe some of this ignorance can be traced to the difficulty K–12 biology teachers and students have concentrating on the topic because of all the screaming from people who believe *The Flintstones* is an animated documentary series. Regard-less, the goal of this chapter requires grasping the basic mechanics of evolution, information that many Americans lamentably lack. What is that goal? In the previous five chapters, we presented the impressive and diverse empirical evi-dence that liberals and conservatives are different in all sorts of ways, from tastes to genes and from personality to physiology. The question we now tackle is why these differences exist. Why are such varied political predispositions present in so many places and at so many times? Whence the deep-seated varia-tion uncovered in the empirical research, variation that runs to the very core of our biological beings? Answers require understanding the theory and accept-ing the fact of evolution because it can explain not only the reason a species has the traits it does but also the reason there is so much within-species variation in those traits.

The Anteater's Snout

Anteaters are unusual looking creatures with extended, gently curved snouts; extremely long, thin tongues; and powerful, digging claws. (See figure 8.1, below.) The latter are used to tear into ant mounds; anteaters then insert their supersized snouts and tongues into the disturbed mound in order to snarf up ants. With a little poetic license, these literally named critters can illustrate all the basic elements of natural selection.[3] If ants are plentiful and close to the surface, snout and tongue length are not crucial to an anteater's survival. If ants are scattered near the top of their mounds, then extended snouts and long tongues are not prerequisites for getting a good lunch. In technical terms, an extra-long snout and tongue package in such an environment confers no fitness advantage to an anteater, meaning those traits do not contribute to survival or reproductive opportunities. If this is the scenario, the genes involved in snout and tongue length, setting aside random variation, are likely to remain the same from generation to generation. Some anteaters will have longer snouts and tongues than others but health and reproductive success will not correlate with these differences.[4]

Now let's suppose the ants wise up. Tired of their brethren being ravaged by such massive, comical-looking creatures, ants begin to spend more time below the surface and get really good at sensing when an anteater is about to lay waste to their mound, scurrying even further underground whenever one of the beasts is poised to set about its business. Anteaters with long snouts and tongues are not much affected by the ants' behavioral change, but anteaters

Figure 8.1   Picture of an Anteater.

with modest snouts and tongues quickly find themselves eating fewer and fewer ants because most of the ants plunge out of their reach. Size now matters. In such an environment, anteaters blessed with long snouts and tongues will grow large and healthy; those with short snouts and tongues will be hungry and eventually malnourished. Variations in reproductive opportunities and success will mirror snout/tongue length. Since anteaters with long snouts and tongues have different combinations of genetic alleles (remember an "allele" refers to a specific version of a gene) than anteaters with short snouts and tongues, the proportion of alleles leading to long snouts and tongues will increase due to the higher reproductive rates of long-snouted anteaters. From one generation to the next, these changes may be too subtle to notice, but they will accumulate and with sufficient time, the genetics of anteaters as a population will have changed merely as a result of the differential reproductive rates of long-snouted as opposed to short-snouted anteaters.

Note that if a massive boulder fell on a young anteater's malleable snout, causing great pain in the short run but an unusually elongated snout in the long run, that particular anteater would be able to eat many ants, be healthy, and probably produce lots of offspring. The situation for those offspring, however, would not be nearly so favorable. They would not have the alleles that would give them long snouts and, assuming no boulder crashed down on them as it did on their parent, would be left with short snouts, empty tummies, and fewer opportunities to pass their alleles to offspring.

With the help of the anteaters, we have now covered the three conditions necessary for natural selection to occur. Pre-existing variation is the first. If before the ants realized the advantages of going subterranean, all anteaters had the same genetically derived snout length, the ants' behavioral alteration would not have instigated any change in the physiology and genetics of anteaters. Darwinian natural selection does not work unless variation exists. Only if alleles at the pertinent loci occasionally vary from organism to organism is it possible for one allele to outperform another (in terms of reproductive success), and this possibility can only come to fruition if the second factor applies. This second factor is known as differential reproduction and it simply means that those anteaters possessing a certain allele (for example, one encouraging longer snouts) are more likely to survive and to produce offspring than those possessing other alleles. Finally, the traits that are helpful to survival and reproduction must be genetically and not environmentally produced. The anteater swinging the long pipe solely because of the accident with the boulder is not a player in the context of real, sustained, Darwinian natural selection. In sum, variation, differential reproduction, and trait heritability, when mixed with the passage of sufficient time, is all that it takes for "humans to develop from earlier animals."

Seen in this light, it is difficult to imagine how evolution could not occur and evidence that it does occur is ridiculously easy to find. Species modification as a result of natural, not to mention artificial, selection is all around us and is anything but "just a theory." The reason you should not overuse antibiotics is evolution, which, in certain situations, selects antibiotic-resistant bacteria just as, in certain situations, it selects long snouts in anteaters. Antibiotics are designed to kill harmful bacteria but all it takes are a few bacteria that, because of their unusual genetic makeup, are not affected by the antibiotic and before you know it, the bacteria with the previously more novel genetic profile are all over the place and the once more common bacteria—the ones vulnerable to antibiotics—are relatively less numerous. You now have a so-called "super-bug," which isn't really super at all, just genetically different enough to survive in the face of common antibiotics.

This descent with modification can be seen in countless other places, including the shifting colors of moths in response to environmental changes such as industrialization (soot-colored wings suddenly became useful for moths eager to blend in and not be seen by predators).[5] An important point here is that evolution is driven by the environment; antibiotics, soot-darkened backgrounds, and behavioral changes in a main food source (such as ants) are all examples of environmental changes triggering Darwinian evolution. Nature and nurture are not fundamentally different sources of change, but instead are inextricably linked. As Gary Marcus points out, though genetic influences are often called "hardwired," a more appropriate metaphor is "firmware," something that is programmed at the factory but always updatable.[6]

Frustratingly, people still argue vehemently over the relevance of nature versus nurture. Yet both influences are rooted in response to environmental changes; the only difference is how fast they occur. Darwinian evolution (nature) takes several generations and is reflected in genetic patterns. Other responses to the environment (nurture) occur much more quickly, over a few months or perhaps merely as much time as it takes an organism to be conditioned or socialized. Nature (in the form of genetic expression) is more affected by the environment than people realize, just as nurture (in the form of socialization, learning, and other environmental inputs) is more conditioned by genetics. The nature-nurture distinction is just not that distinct.

Variation Is the Spice of Life

Standard evolutionary reasoning easily accounts for anteaters' lengthy snouts, beavers' sharp teeth, giraffes' long necks, and the keen vision of eagles. Over generations, poorly endowed anteaters, beavers prone to gum decay, short-necked giraffes, and nearsighted eagles would face severe fitness disadvantages. As a result, the genetic alleles associated with these traits would become rarer in the population. Even so, not all eagles see equally well, and beavers' dental plates are a long way from identical. Why the intraspecies variation in so many different traits? Without breaking a sweat, Darwinian evolution can explain the existence of adaptive traits that characterize a species: trunks for elephants, echolocation for bats, and sophisticated social communication for human beings. But what about the differences that persist within a species? Why do these exist? The answers may help to explain why differences in political predispositions exist.

We've stressed throughout this book that the sources of physical and behavioral variations across organisms of the same species are not always genetic. The environment also shapes predispositions. Still, the extent and consequences of genetic variation are worth examining in depth. Just why are the genes of one person so different from those of another? Shouldn't the alleles conducive to traits facilitating survival and reproduction at some point come to fixation—in other words, come to be present in all humans? As the great evolutionary biologist Ronald Fisher said, "[N]atural selection is a process that eliminates variation."[7] In light of this fact, the substantial degree of human genetic variation is both noteworthy and puzzling.

A favorite statistic of those wishing to minimize human genetic variation is that 98 percent of the genome is the same across all humans. That is probably an overestimate, but let's go with it. The human genome consists of about 3.2 billion nucleotide base pairs (and that is just counting one of each pair of our chromosomes). Even if variations exist in only the remaining 2 percent of the genome, that's still 64 million base pairs—and change in a very small number of base pairs can dramatically alter behavior, physique, and health. In short, there is plenty of very meaningful genetic variation in our DNA.

Consider Huntington's disease. Symptoms typically become noticeable between the ages of 35 and 45, and are horrible. They progress from loss of balance to loss of muscle control to loss of the ability to take care of oneself to loss of sanity to premature loss of life. This is all determined by a nucleotide sequence on the short arm of Chromosome 4. Environmental factors are irrelevant. You can exercise, avoid caffeine, floss daily, meditate, and none of it will make a difference; if your genetics are wrong you will be stricken. In this case what separates right and wrong is about 15 nucleotides. While 64 million out of 3.2 billion (2 percent) may or may not seem like much, it only takes 15 of 64 million (a microscopic fraction of a percent) to shift a vibrant, lucid, healthy, and long-lived human to a twitching, hallucinating, deeply depressed, and prematurely dead one. Indeed, in some cases variation in a single nucleotide can cause dramatic differences in our characteristics and traits.[8] It just does not take much genetic tinkering to meaningfully alter the neurotransmitter systems that help to shape an individual's behavior and health. DNA provides approximately 64 million opportunities for variation from person to person.

We do not just share major portions of our genomes with other humans, but with all carbon-based life forms. Humans share an estimated 95 percent of their DNA with chimps and 60 percent with fruit flies. Put in this context, the 98 percent we share with other actual humans gives a misleading notion of how much we are the same. The truth is that individual human genetic variation is substantial. Many people are uncomfortable with this fact, often because they believe it opens the possibility that various ethnic, racial, and gender groups are genetically different. This concern is overblown. Within-group variation usually dwarfs variation between groups. This means that while genetic variation marks humans as a whole, groups of humans are not particularly genetically distinct. For example, take any two racial groups—say whites and African Americans living in the Southern United States—and you will find much more genetic variation in those two groups than between them. Indeed, the reality of individual-level genetic variation not only fails to support the unsavory ideas of racists, it actively contradicts them. It means Hitler's Nazi group ideal was in reality a genetic mish-mash that overlapped significantly with the genetic variation found in, say, Ashkenazi Jews. So much for Aryan racial superiority.

Even if between-group differences were as large as some apparently fear, normative preference should not blinker us from empirical reality. Population geneticists Gregory Cochran and Henry Harpending put it well: "There is a tradition of caution that approaches self-censorship in discussions of human biological diversity."[9] Avoiding investigation because we are worried about what we might find is a tactic so corrosive to the scientific process and the broader search for truth that it should be adopted only in the most extreme circumstances. And it is not as if the lack of knowledge in this area has proven to be a big boon to politics. The prevalent fiction that there are no meaningful behaviorally relevant genetic differences certainly has not prevented some people (and groups, for that matter) from being treated shabbily or worse. The truth of the matter is that important genomic differences exist across people. This being established, the question becomes not if we should pretend this is not the case but rather why this variation exists.

Human Natures?

A central issue in controversies over the degree and source of human genetic variation is whether this variation is useful (adaptive) or merely occurs for random or at least nonbeneficial reasons. As discussed briefly in Chapter 3, proponents of evolutionary psychology minimize the extent of human genetic variation; they believe each species has a relatively universal genetic architecture that allows individual organisms the flexibility to adapt to their particular environment. For example, compared to girls reared in stable homes, girls reared in less stable family environments marked by divorce and frequent changes of caretaker reach menarche sooner, have earlier and more frequent sexual encounters, and tend to get pregnant at a younger age.[10] There is an evolutionary logic here: Have offspring as early as possible because the precarious environment does not guarantee that there will be opportunities later. The more pertinent point is that the genetic architecture of each girl probably is not specifically designed to reach menarche at a certain time but is designed instead to provide flexibility so that each girl's manifested traits and behaviors can be tailored to her environment. This is the kind of logic that leads evolutionary psychologists to discount genetic variations as explainers of behavioral differences across people.[11]

For example, leading evolutionary psychologists John Tooby and Leda Cosmides famously critiqued the notion that genetic variation leads to distinct personality traits, arguing instead that different personality types are probably the result of "genetic noise" or environmental forces of the sort that allow menarche to be reached at different ages for different girls.[12] While admitting that some variation could slip in the genetic backdoor by piggybacking on a genetically derived trait that *is* variable (probably something to do with the immune system), they insist that variations in personality traits "cannot, in principle, be coded for by suites of genes that differ from person to person."[13] One of their key arguments is that sexual reproduction has the effect of shuffling the genetic cards, so even if one parent was extroverted because of a particular configuration of many genetic alleles, offspring would inherit a different mix of relevant genes from that parent as well as from that parent's mate. In short, they assert that sexual reproduction prevents complicated traits such as personality and, presumably, political predispositions from being heritable and potentially adaptive. The only meaningful genetically based differences across humans they tend to recognize are those between males and females.

This argument is not entirely consistent with empirical observation. Distinct, stable, heritable phenotypes (traits) within a species are sometimes called morphs, and evolutionary psychology seems to argue that behavioral morphs cannot exist in sexually reproducing organisms. Yet they clearly do. Many morphs have already made cameos in this book, including behavioral differences across breeds of dogs and the startling genetically based variations in the behavior of Belyaev's silver fox. Scientists working with different strains of mice laugh at the claim that behavioral morphs cannot exist in sexually reproducing species. In mice—in fact, throughout the animal world—complex behaviors such as personality traits are found to be stable and heritable. Birds pass behavioral tendencies, such as curiosity toward strange objects, to their offspring and do so even when the offspring (thanks to incubation cages) have never seen their parents or any other birds. This is clearly innate, not learned, behavior. Humans are no different; twin studies consistently report substantial heritability coefficients for personality traits as well as for political temperament. This all suggests that personality and political predispositions, while complex, do not require a precise configuration of genetic alleles, but more likely an accumulation of relatively independent, additive influences.

Evolutionary psychology's skepticism of genetically derived behavioral morphs like personality traits is also based in its tendency to assume genetic structure was formed in the Pleistocene (basically, Stone Age times) and has not had sufficient time to change and diversify since. This claim is vigorously challenged by population geneticists Cochran and Harpending, who see the insufficient time argument as simply "incorrect."[14] They document several cases of adaptive genetic changes occurring in the space of just a few thousand years.[15] Examples documented so far include lactose intolerance, language capabilities, and intelligence. Genes relevant to these and other traits give clear evidence of being under recent selection pressure and of undergoing change within the compass of recorded history. This new vision of the pace of genetic change provides a different spin on the nature of human genetic variation. If change can occur that quickly, the existence of variation is much less surprising. Rapid genetic response to environmental change means all human lines are in the process of moving toward some goal that almost certainly will not be reached before the environment changes again, leading to new selection pressures.

Consequently, we question the assertion that "the hypothesis of different human natures is incorrect."[16] There is strong evidence that humans are remarkably varied genetically and behaviorally and that these different predispositions could indeed be viewed as different human natures. In short, both the architecture and the resultant behaviors are far from universal. But what causes these differences? An entire array of possible reasons for genetic variation has been suggested by geneticists and others, and we now describe several.

The Golden Mean

One possible cause of genetic and behavioral variation is what might be called the heterozygote's advantage. Many species are haploid organisms, meaning they have unpaired chromosomes, probably because they reproduce by dividing rather than by having sex. In contrast, humans are diploid organisms, so we have matched chromosomes that form pairs (22 plus the sex chromosomes, to be precise). One of each pair is passed along from mom and one from dad. Each chromosome contains many genes, and some are the same for everybody while others vary substantially from one person to another. It is therefore possible to inherit distinct versions of these varying or "polymorphic" genes from each parent. To illustrate, let's take a simple case where there are only two versions of a gene. Alleles are sometimes indicated by a capital and a lowercase letter (A and a, for example). So depending on what you get from mom and dad you could end up with one of three different genetic combinations: AA, Aa, or aa. Maybe the "A" allele gives people more of something—a growth hormone, let's say—in which case AA's would be taller, aa's shorter, and Aa's somewhere in between, on average. AA's and aa's are homozygous and Aa's are heterozygous.

In many cases, the heterozygous rather than homozygous genotype is advantageous. Perhaps the most cited example of heterozygote advantage pertains to sickle cell anemia, a serious condition where red blood cells take an unusually irregular shape that hinders blood flow. The key gene linked to this disorder is on the short arm of Chromosome 11. If the problematic allele (a) is inherited from both mom and dad, the offspring will have sickle cell, immediately raising the question of why this allele persists in the gene pool. The answer lies in the traits associated with the other genotypes. It might be thought that AA and Aa would be equally adaptive since neither leads to sickle cell anemia (which, as a recessive trait, requires both versions to be "a"), but this is not the case. In certain climates the A allele leaves people more vulnerable to falciparum malaria. So aa genotypes lead to sickle cell anemia but provide some immunity to a common form of malaria. AA genotypes do not lead to sickle cell but are more susceptible to malaria. Thus, it is the heterozygous genotype that is the most desirable.[17] It does not lead to classic sickle cell anemia but still helps to fight off malaria, retaining the good features of the AA and aa genotypes without the bad.

Similar heterozygote advantages have been suggested for intelligence. Cochran and Harpending describe a gene related to sphingolipids, which facilitate the growth of dendrites that allow connections among neurons in the brain. They suggest that individuals with the AA genotype may have a relative paucity of connections, possibly resulting in modest cognitive abilities. The aa genotype, however, is much worse; it can lead to conditions such as Tay-Sachs, a disease that causes progressive deterioration of mental abilities. Cochran and Harpending speculate that the heterozygous genotype may be advantageous in bringing a healthy set of connections that facilitates cognitive ability without leading to neurological disorders.[18]

Heterozygous genotypes that confer these sorts of benefits automatically insure genetic variation in a population because neither the AA nor aa homozygous genotypes are likely to become predominant regardless of selection pressures. Simple Mendelian logic suggests that if one Aa individual mates with another Aa individual and the couple produces four offspring, the odds are that two will be Aa, one will be AA, and one will be aa. In other words, situations in which the heterozygous genotype is advantageous are founts of genetic variation and may be one reason humans are genetically diverse.

The Advantage of Being Unusual

Another potential explanation for persistent human genetic variation is what's known as frequency dependence. The basic logic of frequency dependence was laid out in evolutionary game theory, typically using something called a hawk-dove conceit,[19] with the basic story being that aggressive (hawklike) behavior works, evolutionarily speaking, only when most organisms are dovish. If there are many hawks, the doves will take precautions and the hawk behavioral strategy will no longer be as effective.

Sociopaths are another example of frequency dependence. Homo sapiens are not as fast, fierce, or strong as other species, but our sociality allows us to out-compete these species to the point that we worry about some of them going extinct. Generally we work well with other humans, dividing labor and cooperating in a way that creates something more than the individual parts. This is possible because humans have the knack of quickly sizing people up and trusting those who deserve to be trusted. This basic sociality is exploited by sociopaths. Sociopaths can get people to trust them but, unlike most of us, have absolutely no conscience and seem to delight in exploiting others. When their shtick is about to wear thin, they move to another group where they can repeat their conniving successes. Heightened social skills and no compunction is a powerful combination of forces to use against trusting fellow human beings. If sociopaths can so successfully exploit other humans, why are they so rare? The answer is almost certainly that sociopathy is a frequency-dependent phenotype. In other words, it is a successful evolutionary strategy only if the proportions are right. If there are too many sociopaths, nonsociopaths will be less trusting, undermining the very sociality that sociopaths exploit. In other words, it is the rarity of sociopathy that allows it to work.

Another interesting example of frequency dependence is handedness. Being left-handed or right-handed is heritable and, though there is modest cultural variation, the percentage of left-handers has remained roughly constant since the Stone Age at 10–13 percent of the population.[20] What explains the persistence of left-handers, especially when they are at something of a fitness disadvantage (data show left-handers are generally shorter, lighter, older at puberty, and have a lower life expectancy).[21] There must be a compensating fitness advantage elsewhere. The leading theory about that advantage is based on personal combat and frequency dependence. Combat has been important to humans throughout our history. Losers in combat often are injured or dead; winners gain spoils, prestige, social rank, and reproductive opportunities. What does this have to do with handedness? Right-handers are accustomed to fighting right-handers because they are more numerous. When they find themselves fighting a southpaw, the angles and approach strategies are different and this may shift the advantage to the left-handers. Support for this theory is provided by the generally higher proportion of left-handers in interactive (tennis, fencing, boxing, cricket) but not noninteractive (gymnastics, darts, bowling, snooker) sports and by the fact that left-handedness is significantly more common among (aggressive-interaction prone) males than females.[22] If left-handers become more numerous, the theory goes, they lose their advantage because right-handers would have more experience fighting left-handers. Whether or not this particular explanation is on target, unless the alleles leading to left-handedness are linked to other alleles that are useful, left-handedness must bring some type of fitness advantage that counterbalances the fitness disadvantages.

What this means is that success in large groups can be achieved with atypical phenotypes but only as long as not too many have these traits. If those traits are genetically influenced, the result will be genetic variation. With the right frequencies, being left-handed or a sociopath can be a successful evolutionary strategy, which means genetic diversity relevant to those phenotypes. A related concept is niche-filling. Organisms with distinct genetic predispositions can gravitate toward particular strategies. Small black bears can reach their paws into tight places to extract honey; big black bears can use their massive, powerful paws to snag rapidly swimming 50-pound salmon. If the food source of black bears was less diverse, the genotypes of black bears would likely be more consistent across bears. Different niches leading to perfectly respectable levels of fitness foster genetic variation. Some niches may be limited in the number of organisms they can sustain, resulting in another form of frequency dependence.

Ongoing Selection, Byproducts, Varying Ancestral Environments, and Randomness

A more obvious source of genetic variation is that we are catching evolution somewhere in mid-process. Long anteater snouts are immediately beneficial when the ants begin hanging out further underground but this does not mean the genetics supporting long snouts (and tongues) will turn on a dime. For generations and generations, alleles relevant to snout length will be mixed and therefore so will snout length (some long, some short, and many in the middle). Only gradually will the long-snout faction become more numerous and it would take many, many generations to reach a stage where all anteaters had extremely long snouts. Until that time, genetic variation will be the order of the day.

Another possible cause of variation is that we are not seeing evolution—at least not natural selection of adaptive traits—at all. It is a mistake to assume that every trait or variation in every trait is adaptive. Some nonadaptive traits may be related to a trait that is adaptive and piggyback on that trait for the evolutionary ride. One recently suggested example involves religion. Why are humans prone to holding religious beliefs and why do these beliefs vary so much? Some scholars argue religion is a byproduct of the adaptive tendency to attribute intentionality to objects and forces of nature.[23] Assuming intentionality even when there is none probably creates an evolutionary advantage compared to assuming no intentionality, even though it is sometimes there. This is because being very good at detecting patterns (even when they are not there) helps to create a default way of thinking that allows us to make sense of the world and make quick choices about how best to survive and prosper in it. Being very good at detecting randomness—in other words, being able to recognize quickly the absence of intentionality—offers no such consistent advantage. As such, selection pressures push toward those who see intention-ality where there is none. An unsurprising byproduct of these evolutionary pressures would be large numbers of people who believe in God, Gods, spir-its, ghosts, angels, and demons. This trait could be common even though the underlying genetic proclivities and variations that foster religious beliefs are not directly adaptive.

An even better example is reflected in recent research on depression. It turns out that there is a connection between depression and immune-relevant inflammation. A logical assumption is that those individuals who are prone to depression have bodily responses to their depressed mood. Undoubtedly this happens, but the reverse may also occur—inflammation comes first and the depression is a byproduct. Andrew Miller and his colleagues document that depressed people have higher levels of inflammation even if they are not fight-ing disease.[24] This can be good, especially in children, because the body is ready for a surprise invasion. Primed immune response is obviously advantageous in pathogen-heavy environments. But a byproduct of this heightened immune readiness and its accompanying mild inflammation is the release of cytokines, a type of signaling molecule. These cytokines are active in the brain and correlate with depressive symptoms in a significant subset of the population. The current thinking is that depression does not cause these immune-relevant responses but rather is a byproduct of them.[25] The tradeoffs between the fitness disadvan-tages of depression and the fitness advantages of a vigilant immune system thus might further contribute to genetic variation.

Genetic variation might also exist because evolution can happen more than once; different environments can create different selection pressures on the same species at roughly the same time. The standard evolutionary psychology paradigm, which assumes that a stable and monolithic environment of evolu-tionary adaptiveness (EEA) locked humans into the traits they have today, is increasingly under attack.[26] Pleistocene epochs were not uniformly stable and substantial variation probably existed from region to region and even from group to group. If environments varied and there was some stability over time within these micro-environments, then evolutionary pressures were different in each. Elizabeth Cashdan documents the differences between hunter-gath-erer societies that have long lived in conditions of plenty and those surviving for generations in conditions of scarcity.[27] Not surprisingly, attitudes toward egalitarianism and related concepts are quite different. Some genetic and trait variation observed in modern humans might be traceable to just these sorts of differences in ancient micro-environments.

Sometimes genetic variation occurs for no particular reason at all; it just happens. This is called genetic drift, and it contributes to the diversity of life. Indeed, if no errors (random mutations) ever occurred in DNA replication, the first, simplest life form could never have evolved into anything more complex. Particularly in new and small populations, genetic drift can play a major role in shaping the species and accounting for genetic diversity.

The Borg

The bubonic plague raced across Europe from 1348 to 1350, killing an esti-mated 50 percent of the continent's population. Its cause was the Yersinia pestis bacterium, carried by fleas riding on stowaway rats in merchant ships and Silk Road carts.[28] The plague was scarily lethal, so why did one out of two Europeans survive the "black death"? Because, thankfully, not all human immune systems are the same. Immune systems make it easy to see the advan-tage of genetic diversity, but note that this advantage accrues to groups rather than individuals. If every member of a species is identically vulnerable to the same pathogen, that pathogen can be an agent of extinction. If organisms vary in their immune strengths and weaknesses, on the other hand, any given pathogen might knock back the species, but it won't wipe it out. This is why the plague set population growth in Europe back 150 years but did not lead to the end of humanity.

The shift in focus from the individual to the group is worth highlighting. It does not need to get to the level of the Borg, the Star Trek aliens bent on mold-ing everything into the "hive mind" of the collective. This shift in perspective, though, not only involves an ongoing debate about the way evolution works but also has big implications for the application of evolutionary theory to politics. Most evolutionary biologists are still more comfortable viewing natural selec-tion as working on the traits of individuals rather than the traits of groups or species; however, the minority view that evolution could work on groups as well as individuals—what is known as group or multilevel selection—is gaining strength. A powerful voice supporting this perspective is biologist E. O. Wilson, who challenges conventional arguments by asserting that individual organisms sometimes sacrifice for other organisms not only because they might be part of the same extended family and therefore share some genetic heritage (kin selection) but simply because, related or not, they are part of the same group and will benefit from that group being strong.[29]

Group selection is an idea with empirical backing from animal selection studies. To take an example highlighted by multilevel selection advocates Eliot Sober and David Sloan Wilson, we turn to a chicken laying an egg.[30] This may seem like a solitary act but that is not entirely correct, as the poultry business discovered the hard way. When the most prolific egg layers (variation in the pace of egg laying is a heritable trait) are identified and combined into a supposedly super-fecund flock, production drops dramatically. How can this be? Apparently because the most prolific egg layers also tend to be somewhat aggressive prima donnas and if you put a bunch of aggressive prima donnas together, the result is not a healthy, productive unit. Given this, poultry husbandry began to consider chicken societies rather than individual chickens. By scoring the egg-laying productivity of groups of chickens and using the best groups for future breeding, egg productivity increased 160 percent in just six generations.[31] The key was to find chickens that fit together well as a group and not just to throw together the most productive individual chickens.

Some biologists acknowledge this sort of evidence and conclude that evolution might work on levels such as groups. Others remain skeptical, though without group selection it is difficult to explain the sometimes unusual levels of altruism observed in humans and many other organisms.[32] If evolution does work at the group level, this would have important implications for genetic diversity within those groups. Specifically, we would expect to see quite a bit of it. Think about it this way. Chicken flocks are more successful when they are populated not exclusively by shrews that crank out eggs at a remarkable rate but by rich mixtures of personality types. This form of selection pressure is almost a division of labor, with some hens laying eggs like crazy and others laying fewer eggs but keeping the group vibe pleasant. Hens can thus contribute to group success by playing any of a variety of genetically molded roles. Group selection is quite a different way of thinking about evolution and makes it easy to see the advantages of genetic diversity.

Political Variation: Why Are There Liberals and Conservatives?

With this range of possibilities for genetic variation in mind, let's apply the basic ideas to the key focus of the book: the fact that political variation (a heritable trait, as we have seen) is extraordinary in virtually all societies for which data are available. There are no known cultures in which everyone agrees on politics, just as there are no known cultures in which everyone has identical personalities. A mass society with apparent universal agreement on every political issue is almost certainly a mass society where free expression is unknown. How does evolution account for the existence of liberals and conservatives, moderates and extremists, the left and the right, progressives and curmudgeons, "traditional warriors" and "new villagers,"[33] and "ethnocentric hawks" and "empathic doves"?[34] Answers are unavoidably speculative—unlike in previous chapters, here we cannot support our arguments with a mound of empirical evidence. What we can do, though, is use what we have learned about evolution and genetic variation to offer plausible explanations of the origins of variations in political predispositions.

A common question we get when people learn that our research deals with the biological differences of liberals and conservatives is whether one ideology is "more evolved" than the other. Typically, liberals are eager to be told that conservatives are some type of antediluvian life form and conservatives are equally eager to learn that liberals are at odds with the natural order. Sorry to disappoint, but this type of question is silly. The truth of the matter is that concepts such as "more" or "less" evolved are nonsensical. Evolution is the process of species adapting to their environments and, because the environment itself is a moving target, the process is never ending. Evolution is not a destination but a temporary and sometimes lagging accommodation to environmental realities that existed at a certain time. If the environment shifts again, evolution will begin to move in a different direction, so no genetically based political predisposition is rightly viewed as more or less evolved.

Scholars differ on the related issue of which political predisposition is more "natural." Some assert that the great explosion of human culture some tens of thousands of years ago created the basis for two politically very different types of human. The first ("traditional warriors" or conservatives) reflects the state of the species prior to the great cultural flowering and the latter ("new villagers" or liberals) reflects the status after. The implication is that conservatives are somewhat out of step with current sensibilities.[35] Others have exactly the opposite view, asserting that Darwinian behavior is essentially self interested (this perspective obviously downplays group selection) and that it is liberals, with their absurd notions of the perfectibility of mankind, faith in international tribunals and governmental programs, and dislike of competition, who are out of step with the great sweep of history.[36] So a case can be made either way. In truth, the issue of which ideology is more natural (and thus the wave of the future) is as big a philosophical dead end as arguments over which ideology is more evolved.

A more interesting issue is the particular type of environment for which liberals as opposed to conservatives are best suited. What can be said about this matter? It seems relatively uncontroversial to suggest that times were tougher, more dangerous, and more Hobbesian in our distant evolutionary past—Hobbes being famous for, among other things, saying life is "solitary, poor, nasty, brutish and short." Warfare and homicides claimed a startling number of lives back then relative to the current day—forensic archeologists' and quantitative ethnographers' best estimates are that perhaps half of all males died at the hands of other males.[37] Focusing purely on deaths in battle, the rate may have dropped from 500 out of 100,000 in pre-agricultural times to 0.3 out of 100,000 in the contemporary era. The probability of being killed by another person has dipped dramatically even off the battlefield and even in the last 750 years. One estimate is that in fourteenth-century England, 24 deaths out of every 100,000 resulted from homicide, but that by the late twentieth century the comparable figure had dropped to less than 1 per 100,000.[38] That is a long way from one out of two males being put in the ground because another male bashed them on the head. The twentieth century, with two world wars and numerous other conflicts, is remembered as particularly bloody; yet if the rates of violent death in prestate societies had applied in the twentieth century, an estimated two billion people would have died—far, far more than actually did.[39] And death at the hands of others was not the only problem in pre-agricultural societies. Death from pestilence and accidents also was much more common given the nature of the times and the absence of sophisticated medical care.

In sum, existence in hunter-gatherer societies prior to the advent of mass agriculture was short and filled with a remarkable range of threats. Selection pressures in such environments would likely favor individuals with higher degrees of negativity bias, who approached novel situations with caution, who were loyal to their group, and who were suspicious of the tribe over the hill. These would be the individuals most likely to avoid danger given that they would be less likely to open themselves up to situations in which they would be vulnerable. Such individuals would be responsive and attentive to threats. Given the evidence presented in the previous chapters, they would also have been the individuals who, in a modern mass polity, would display conservative political predispositions.

Our best guess is that in the rough and tumble of the Pleistocene, individuals who tried new things, opened themselves up to members of other tribes, and had little to no negativity bias were rare—it simply seems a losing long-term strategy in the face of all the dangers swirling about. Social units relatively isolated from threats for long periods of time might have permitted some protoliberals in the mix, but most hunter-gatherer groups would likely have needed to keep a constant eye on the horizon and maybe even on the next hut. These were likely conservative societies in the sense that they did not often make big changes in the way they did things and those genetically inclined to take chances, to go through life marching to their own drummer, were probably selected against.[40] As Jonathan Haidt puts it, we are likely "descendants of successful tribalists, not their more individualistic cousins."[41]

In many respects, with the advent of large-scale agricultural societies around 12,000 years ago, life started on a path to becoming less threat prevalent, though the initial transition from small bands of seminomadic individuals to large, stationary polities was hardly a universal plus. Diseases increased due to living in such close proximity to so many other people, not to mention to so many large, domesticated, and often infectious disease–carrying animals; water and sanitation quality diminished; hierarchies and discrimination became apparent and oppressive; and nutrition declined.[42] Eventually, however, humans adjusted to the new lifestyle, food sources became more predictable, and quality of life increased, though unevenly and with many serious setbacks (e.g., the Dark Ages). Over time, the chances of dying violently declined. Sanitation, medical care, and nutrition eventually improved as well. In such an altered environment, selection pressures for heightened negativity bias, for the tried and true way of doing things, and for deep suspicion of out-groups likely would start to fade.

Such environmental changes would not necessarily mean that openness to out-groups and new experiences suddenly started to be positively selected for. Greater trust of others, exploratory behavior, and a more relaxed orientation toward negative elements of the environment certainly can be beneficial given that they increase the possibilities of trading with other groups, learning from others, and discovering better ways of doing things. Trust has been shown to be remarkably beneficial to societies and is more difficult to display if the prevailing attitude is ethnocentric and fixated on potentially negative consequences.[43] So positive selection pressures for open, trusting, and exploratory orientations might have increased a degree or two, but it is more likely that humanity's shifting social environment merely relaxed the strong pressure to be cautious and attentive to the negative. A logical result of this would be for traits like negativity bias, attitudes toward out-groups, and openness to new or novel experiences to vary more widely than they had in the Pleistocene.

Most people in the developed portions of the world today simply do not have the same constant, life-threatening worries that existed in the distant past. As a consequence, people today can "expand their circle" of social contacts and ethical concerns beyond family and tribe to people far away and perhaps even to animals.[44] Not everyone will take this opportunity, and the absence of strong selection pressures will encourage tremendous variation in genetically influenced predispositions toward what in modern mass-scale societies is called either liberalism or conservatism. Liberalism may thus be viewed as an evolutionary luxury afforded by negative stimuli becoming less prevalent and less deadly. If the environment shifted back to the threat-filled atmosphere of the Pleistocene, positive selection for conservative orientations would reappear and, with sufficient time, become as prevalent as it was then.[45]

The basic evolutionary scenario we are sketching is one where the selection pressures for being "conservative" in social outlook ease off and variation increases. Neither preference for the tried and true nor eagerness to try something novel is necessarily more adaptive, and if a trait is not strongly adaptive, there is likely to be more variation in it. As Tooby and Cosmides put it, "variation tends to occur wherever uniformity is not imposed by selection."[46] This line of thought also squares with an empirical reality that appears fairly regularly in our results as well as in the results of others. Conservatives have clearer tendencies than liberals. To take just one example, conservatives tend to have a negativity bias; liberals sometimes do and sometimes don't. Liberals do not have a positivity bias; their orientation to positive as opposed to negative stimuli is more varied and not as clear as negativity biases are for conservatives. This sometimes bothers conservatives who grumble about liberal academics implying that conservatism (and conservatism alone) is in need of explanation, almost as though it were pathological.[47] In reality, conservatism may be a tighter phenotype than liberalism simply because at one time it was actively selected for.

Do Liberals and Conservatives Need Each Other?

If we stopped here, our evolutionary account would be incomplete because traits not under heavy selection pressure should vary randomly and that is just not the case for political predispositions and the traits associated with them. Some observers contend that political predispositions are genetically based but randomly distributed across a population;[48] twin studies (see the previous chapter) strongly suggest that predispositions are genetically influenced but definitely not random. Political dispositions not only run in families and in patterns, the variation in these predispositions appears to be becoming more patterned and less disorganized all the time. Talk of polarization—where political opinion splits and gravitates to the extreme ends of the ideological spectrum—is everywhere. Even with many people in the middle, the liberal-conservative spectrum seems to divide political battle lines better than ever. If we are correct about the relative absence of selection pressures, why hasn't that lack of selection pressure left us with only random variation in the variables associated with political predispositions? The stability and durability across societies of the split described by Emerson (progressives and traditionalists) and evident in our physiological and psychological data, along with the heritable variation of political views, suggest something more than just the removal of selection pressures for conservatism has occurred.

This is where several scholars have suggested that group selection might play an explanatory role in that distinct advantages may accrue to groups that have a variety of political perspectives just as there is an advantage to groups having a variety of immune systems. A mix of those with and without strong negativity biases, those willing and those unwilling to take risks, and those welcoming and not-so-welcoming to out-groups might make for a stronger social group compared to a group entirely made up of one type or the other. The argument for the group selection of political predispositions has been around for more than 20 years. In 1992, former member of Congress Jim Weaver, whom we met in the previous chapter, noted that "had we all been empathic, we would have been slaughtered by other hominids many thousands of years ago; had we all been violent, aggressive beasts . . . we would not have developed the creative skills that made us the dominant animals on earth."[49] In one of our studies, using somewhat different terms, we speculate that "as loathe as contextualists and absolutists are to admit it, the presence of the other orientation may make a society stronger,"[50] a theme that has been developed more recently by Jonathan Haidt.[51] The argument is that societies with a mix of political types would be better able to adapt to changing environments because they would have some members who were more attendant to defending the in-group and others more eager to engage with out-groups; they would have some members who were more willing to try novel approaches and others more eager to stick up for the old ways. What would be the big downside of this sort of mix? Well, groups such as these would have many political disagreements to resolve and might not be very good at resolving them—in other words, they would look like us.

For this explanation to hold up, a mix of phenotypes needs to be preferable to a standard phenotype flexible enough to flip individuals between protoconservative and protoliberal depending on conditions. In other words, it needs to deal with the universal architecture argument championed by evolutionary psychologists especially because at least some of this flexibility obviously exists. For example, New Yorkers, who include some of the most liberal individuals in the United States, became more conservative after 9/11.[52] Still, this is movement from a pre-existing predisposition or set point. Why do these set points exist at all when total flexibility might solve the same problem—in other words, why aren't we all conservative *and* liberal, switching from one to the other based on signals from the environment? Wouldn't this be better than having distinct groups that don't much like each other? The relative advantage of type as opposed to total flexibility is difficult to test, but some evidence on the matter comes from computer simulations. One study examined the value to virtual societies of two different types of "heroes": those who demonstrate heroism in across-group conflicts (military types) and those who demonstrate heroism in within-group activities (altruists). The study finds the most successful groups are those composed of a different mix of types rather than those peopled by individuals flexible enough to fulfill both roles. This result suggests a benefit for distinct behavioral phenotypes over wholly flexible phenotypes.[53]

The existence of morphs in much of the animal world is further indication that variation in predispositions is not accidental. The division of labor among bees and ants and the simple existence of relatively stable personalities among all sorts of animals suggest there is value in certain kinds of diversity within groups. Just as groups of organisms benefit from a division of labor, they probably also benefit from a division of social and perhaps now political predispositions.

Given the recentness of mass-scale society, the core traits would have to apply to small-group, hunter-gatherer life because people have not been able to meaningfully express their political views for very long. Since the establishment of mass scale polities maybe 10,000–15,000 years ago, democracies have been rare and recent. The brief fling in Athens was an aberration, and as late as 1945 only 20 democracies existed; many of the largest countries from a population point of view (e.g., China) were not democratic then or now. In the history of the world, only a fairly small number of people have ever lived within a mass-scale democracy, so we as a species simply have not had much practice living in this sort of social environment. Even if Cochran and Harpending are correct, and they are, that natural selection can occur much more rapidly than previously thought, selection pressures for a mix of liberals and conservatives in mass polities simply have not had enough time to work.[54] The pressures we describe must have been for diverse social predispositions in small-scale bands, predispositions that later manifested themselves as liberals and conservatives or progressives and traditionalists when mass-scale democracies came on the scene.

We believe that traits such as orientation toward out-groups, openness to new experiences, and a heightened negativity bias fit more naturally with social than economic issues, and we tend to agree with Congressman Weaver that economic positions are typically secondary. He points out that "ethnocentrics do not give a fig for individual rights" and sees the connection between conservatism and free market principles as a relatively recent development. Similarly, he does not view Marxism as connecting to the deeper forces shaping empathics and believes that accounts that do make this connection "totally ignore our biological origins."[55] The deep forces that shape political predispositions likely do not act directly on controversies over the role of government in society (after all, for how long in evolutionary time has the size of government been an issue?) or, relatedly, on controversies over the glories of the free market relative to the social welfare state. But if the issue becomes whether or not to open up a country's social welfare system to recent or future out-group members (that is, immigrants), deeper forces quickly come into play. Economic issues are certainly crucial in modern politics—sometimes the most crucial—but this does not mean fault lines on these issues are as biologically rooted as social issues.

One More Time: Not by Genes Alone

We have been focusing on the role of genetic variation, but remember that genes are not the only cause of political temperament. The obvious relevance of the environment to political orientations suggests that fairly small genetic differences get magnified by environmental forces to create distinct political predispositions. Christopher Jencks's work on reading proficiency provides a good example of how such a process might work.[56] He notes genetic differences in peoples' reading ability but points out that if reading comes easily to someone, he or she is likely to read more books on average, thereby becoming more proficient and drawing encouragement and praise from parents and teachers, which would lead to more reading, and so on. The environment, in short, amplifies the initial, modest genetic variations. Similarly, relatively modest differences in social proclivities could be exacerbated by parents and other influential actors. Individuals with slight tendencies toward caution and tradition might gravitate to others with similar tendencies and in the current era perhaps toward media outlets that sing the same tune. These environmental influences pile on and the result eventually is a full-fledged predisposition.

These developmental and environmental forces sometimes may push against genetic type, which would help explain the occasional major political conversion and may be why converts are often more extreme than those who have been consistently ensconced as a liberal or conservative all along. Going home can be liberating. Neoconservatives provide several examples of this. Jeane Kirkpatrick is best remembered as a right-leaning Cold War warrior;

she served in Ronald Reagan's administration, took strident anti-Communist views to her position as U.S. ambassador to the United Nations, and formulated the "Kirkpatrick Doctrine," which called for the United States to throw its support behind anti-Communist regimes even if their records on democracy and human rights were, shall we say, questionable. This paragon of the New Right started her political journey out on the left-wing political fringe; as a young adult she was a member of the Young People's Socialist League, which was affiliated with the Socialist Party of America. And let's not forget about all the people in the middle. Moderates could be the result of a mix of cross-cutting environmental and genetic forces, a diminished level of political interest and awareness due to different environmental and genetic forces, or simply an absence of dispositional forces. Just because some people are shaped by these predispositions does not mean everyone has them. In fact, it may be that many people do not but that the people who are predisposed tend to be those who are by far the most vocal. Having a large number of moderates and apoliticos does nothing to diminish the general account being presented and in fact meshes with it perfectly.

Conclusion: Escape Routes and Packing Heat

Oldfield mice are well known for the long burrows they dig, usually with a separate escape tunnel. (See figure 8.2, page 229.) Scientists long suspected a genetic component to this behavior, since even after generations of living in metal cages, one of these mice let into the wild will dig a burrow. Still, the burrows vary both in length and in whether or not they include an escape tunnel. What explains this intraspecies variation? Three scientists at Harvard used foam molds to determine the exact dimensions of each burrow and sought to correlate these burrow dimensions with genetic variation. They discovered three regions of the mouse genome where genetic variation correlated with the length of burrow. Each of these regions has an independent and roughly equal effect of adding about 3 centimeters to burrow length. They also identified one particular gene that seems to govern whether a mouse will add an escape tunnel to its burrow. If the mouse has at least one dominant allele at this locus, it is 30 percent more likely to build a getaway route, and this escape route–relevant gene is entirely separate from the genetic regions relevant to the length of the burrow.[57]

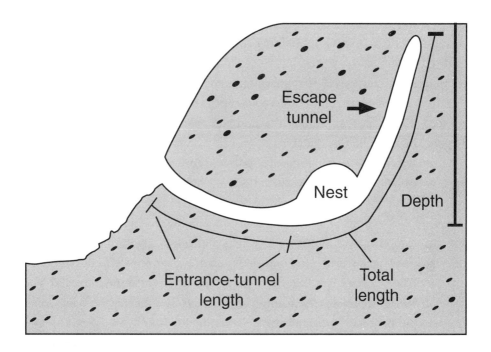

Figure 8.2 Picture of a Mouse and a Mouse Burrow

These findings are quite extraordinary. Previously, allelic variation was connected to physical traits and conditions or to simple behaviors related to eating or reproduction. Digging tunnels is complex behavior; size, location, contour, and route must all be sorted out. The influential role of a surprisingly small number of genes in this relatively complicated behavior came as a bit of a shock. The fact that the decision to build an escape tunnel or not is so closely tied to one gene flies against arguments that complex behavior can never be connected to genes in a way that is tractable.

Is human decision-making about politics more complex than mouse decision-making about burrow construction? We hope so, though the evidence is mixed.[58] Humans certainly can incorporate much more information, but we too have inclinations shaping our decision-making. In many respects, adding an escape route to a burrow is a means of providing security. If a snake approaches the main entry, the mouse (and family) can deal with the threat by escaping. How different is this from the strong desire of some humans to be protected from threats to their home and family? The mechanisms by which humans pursue security might involve agitating for "stand your ground" laws that allow the use of deadly force if an unlawful threat is perceived, supporting higher defense and law enforcement budgets, and fighting doggedly for the right to be armed to the teeth, but the underlying instinct could well be comparable to a mouse building protection into its burrow. We will likely never be able to understand let alone predict with anything approaching complete accuracy the decisions people make, but if the "burrowing" research is any indication, the forces underlying these decisions are not as complicated as generally imagined.

The ability of one gene to exert a powerful influence on a complex, multifaceted behavior such as a mouse digging an escape tunnel or a fruit fly performing a variegated pre-mating ritual suggests we need to think more creatively about the ways genes can affect behavior. The presence or absence of one protein (the product of a gene) should not be able to shape such complex behaviors. It almost seems as though neuronal arrangements of some sort are in place for reasons not fully understood (epigenetics?) and that the protein produced by the key gene either kicks this arrangement into gear or not. In this fashion, one gene may shape complex behaviors more than previous seemed possible. Our prediction is that complex behaviors such as tunnel digging and political predispositions will serve as the platform for discovery of a new style of trans-mission—neither learned nor traditionally genetic.

The modularity in mice's decisions about their burrows—that is, the fact that decisions about escape routes are separate from decisions about the length of the main burrow—is also useful for thinking about decision-making in the realm of politics and could help to explain the "modularity" or different ele-ments of a larger ideology, such as positions on economic as opposed to social issues. Remember, evolutionary psychology suggests there cannot be behav-ioral morphs in sexually reproducing organisms like humans and mice because complex behaviors must emanate from elaborate interactions of genes and the environment, so if one gene is altered the whole house of cards comes tum-bling down. The conclusion is that precise configurations cannot be passed along from generation to generation, in other words, are not heritable. Maybe so, but it may also be the case that behaviors such as complex burrowing are not the result of highly specific configurations of large numbers of genes but rather are attributable to a modest number of independently operating alleles. The three distinct regions of the mouse genome correlating with burrow length affect complex behaviors without needing to be configured in a precise fash-ion. We suspect the forces shaping human political decision-making work the same way. Alleles at numerous sites can push us toward a high negativity bias (wouldn't it be nice to know if the mice that dig escape routes also respond more to negative stimuli, such as an image of a snake, than do the mice that do not dig escape routes?); suspicion of outsiders; and a reluctance to try new behav-iors when traditional options are available (are mice that fear novel objects also more likely to dig escape routes?).

If we are correct, rather than a single demanding configuration, there are many different genetic and biological ways to be politically conservative (or lib-eral), and many different and separate elements of being either conservative or liberal. We believe this hypothesis about multiple routes is testable. Researchers tend to take one psychological or physiological concept at a time and determine if it is related to political predispositions. As we have seen, political tempera-ment has been found to correlate with skin conductance response to negative images, with psychological preferences for closure and certainty, with particu-lar moral foundations, with specific tastes in food and art and leisure pursuits, with contours of the amygdala, with variations in the kinds of stimuli to which

attention is paid, with variations in neural patterns as a result of seeing something unexpected, and with distinct levels of exploratory behavior, to name just
a few. Yet little research has attempted to determine if these various correlates
of political predispositions are correlated with each other. In other words, are
those people who prefer their poetry to rhyme the same ones whose skin conductance elevates when they see images of wrecked cars, vomit, or houses on
fire? If political predispositions are as modular as mouse burrowing behavior,
and we have no reason to believe they are not, these correlates of political orientations are likely to operate fairly independently. The data required to test
this possibility are hard to come by (since information on many variables is
required) but our lab is in the process of conducting preliminary analyses.

Recent research on what are sometimes called "behavioral syndromes" is
clearly relevant to understanding the nature of variation in political predispositions.[59] These studies note the surprisingly high correlations of fairly diverse
phenotypes in several different animal species. For example, in numerous species
variations in foraging behavior are found to correlate with variations in mating
behavior, antipredator behavior, territoriality, and aggression. Why would these
seemingly disparate behaviors correlate? Presumably for the same reason the
seemingly diverse components of the conservative or the liberal phenotype fit
together. The fascinating possibility this raises is that downstream behavioral
syndromes, whether in mice or humans, might stem from modular genetic forces
that are then amplified by (usually) supportive environmental experiences to
create remarkable intraspecies variation in a range of behaviors, including those
pertaining to human mass-scale politics; in other words, to create predispositions.

The empirical work summarized in Chapters 4–7 described the reality of
political predispositions at some length; in this chapter we offered a plausible
account of why such different predispositions exist. The final task before us is a
consideration of the manner in which an understanding of politically relevant
physiological and deep psychological differences might improve human social
life in the early twenty-first century.

Notes

[1] Miller, Scott, and Okamoto, "Public Acceptance of Evolution."
[2] For the record, a surprising number of Americans are willing to accept that many species
     evolve as long as humans are not included in this list. The notion that all other life forms

except humans are subject to natural selection suggests that blatantly irrational human exceptionalism is alive and well.

[3]   Darwin did not use the term "evolution" until it became popular in lay discussion. "Natural selection" was his preferred phrase—or "descent with modification."

[4]   In biological parlance, this is known as Hardy-Weinberg equilibrium.

[5]   Kettlewell, "Darwin's Missing Evidence."

[6]   Marcus, *The Birth of the Mind*, 40.

[7]   Fisher, *The General Theory of Natural Selection*, 2nd ed. See also Tooby and Cosmide, "On the Universality of Human Nature and the Uniqueness of the Individual," 37.

[8]   The condition known as phenylketonuria, or PKU, one of the more common sources of mental retardation, is an example.

[9]   Harpending and Cochran, "In Our Genes."

[10] Quinlan, "Father Absence, Parental Care, and Female Reproductive Development."

[11] Though later research has emphasized that there is a genetic component to age at menarche. See Morris et al., "Family Concordance for Age at Menarche."

[12] Tooby and Cosmides, "On the Universality of Human Nature and the Uniqueness of the Individual."

[13] Ibid., 19. For a slightly different take, see Buss and Greiling, "Adaptive Individual Differences." For yet another perspective, see Figueredo et al., "Evolutionary Theories of Personality."

[14] Harpending and Cochran, "In Our Genes," 10.

[15] Cochran and Harpending, *The 10,000 Year Explosion*.

[16] Tooby and Cosmides, "The Past Explains the Present: Emotional Adaptations and the Structure of Ancestral Environments," 23.

[17] Actually, the heterozygous genotype can lead to sickle cell trait, a milder but still potentially dangerous condition in which the red blood cells under certain circumstances take on the sickle cell shape, usually when exertion is inordinately high (for example, demanding athletic competitions at relatively high altitudes).

[18] Cochran and Harpending, *The 10,000 Year Explosion*.

[19] Smith, *Evolution and the Theory of Games*.

[20] Raymond et al., "Frequency-Dependent Maintenance of Left-Handedness in Humans."

[21] Yeo and Gangestad, "Developmental Origins of Variation in Human Preference."

[22] Raymond et al., "Frequency-Dependent Maintenance of Left-Handedness in Humans."

[23] Pyysiäinen and Hauser, "The Origins of Religion: Evolved Adaptation or By-Product?"

[24] Raison and Miller, "The Evolutionary Significance of Depression in Pathogen Host Defense (PATHOS-D)."

[25] Depression is likely to be maladaptive, though there is a case to be made that it can be adaptive in some circumstances (cut your losses by staying home in bed rather than by banging your head against the wall of a world that is causing you pain).

[26] Cochran and Harpending, *The 10,000 Year Explosion*. See also Buller, *Adapting Minds*.

[27] Cashdan, "Egalitarianism among Hunters and Gatherers."

[28] Kohn, *Encyclopedia of Plague and Pestilence*, 3rd ed.

[29] Nowak et al., "The Evolution of Eusociality."

[30] Sober and Wilson, *Unto Others: The Evolution and Psychology of Unselfish Behavior*.

[31] Muir, "Group Selection for Adaptation to Multiple Hen Cages."

[32] De Waal, *Good Natured: The Origins of Right and Wrong in Humans and Other Animals*.

[33] Haston, *So You Married a Conservative*.

[34] Weaver, *Two Kinds: The Genetic Origin of Conservatives and Liberals*, 5.

[35] Haston, *So You Married a Conservative*.

36 Arnhart, *Darwinian Conservatism*. See also Gabler, "The Weird World of Biopolitics."
37 Daly and Wilson, *Homicide*.
38 Eisner, "Modernization, Self-Control, and Lethal Violence: The Long-Term Dynamics of European Homicide Rates."
39 Keeley, *War before Civilization*. For a full treatment of the decline in violent death over the millennia, see Pinker, *The Better Angels of Our Nature*.
40 This is similar to Haston's speculation in *So You Married a Conservative*.
41 Haidt, *The Righteous Mind*, 163.
42 Cochran and Harpending, *The 10,000 Year Explosion*, 85–90.
43 Robert Putnam, *Making Democracy Work*; and Francis Fukuyama, *Trust: The Social Virtues and the Creation of Prosperity*.
44 Singer, *The Expanding Circle: Ethics, Evolution, and Moral Progress*.
45 Haston sees progressivism as the wave of the future, but this belief seems to presuppose the environment will continue to move in the direction it has been moving.
46 Tooby and Cosmides, "The Past Explains the Present: Emotional Adaptations and the Structure of Ancestral Environments," 58.
47 Will, "Conservative Psychosis."
48 Weaver, *Two Kinds: The Genetic Origin of Conservatives and Liberals*.
49 Ibid., 12.
50 Alford et al., "Are Political Orientations Genetically Transmitted?," 166.
51 Haidt, *The Righteous Mind*.
52 Huddy and Feldman, "Americans Respond Politically to 9/11."
53 Smirnov et al., "Ancestral War and the Evolutionary Origins of 'Heroism.'"
54 Cochran and Harpending, *The 10,000 Year Explosion*.
55 Weaver, *Two Kinds: The Genetic Origin of Conservatives and Liberals*, 5.
56 Jencks, "Heredity, Environment, and Public Policy Reconsidered."
57 Weber et al. "Discrete Genetic Modules Are Responsible for Complex Burrow Evolution in Peromyscus Mice."
58 Antonakis and Dalgas, "Predicting Elections Is Child's Play."
59 Bell, "Future Directions in Behavioral Syndromes Research."

CHAPTER 9

Can Conservaton and Liberalville Survive Together?

[If political attitudes are genetically influenced] it would require nothing less than a revision of our understanding of all of human history, much—if not most—of political science, sociology, anthropology, and psychology, as well as, perhaps, our understanding of what it means to be human.

Evan Charney

Everyone is entitled to their own opinions but they are not entitled to their own facts.

Daniel Patrick Moynihan

Conservaton is, for some people, the perfect place to live. Its neighborhood watch program is vigorous but hardly needed because people are law abiding, not to mention heavily armed. The schools emphasize discipline and respect for authority, and build their curriculums around rule-based instruction like phonics for reading and memorization of formulas for math. Conservaton's similarly designed houses are well maintained, clad in pretty much the same two colors of vinyl siding, and fronted by beautifully manicured lawns. There is a church on nearly every block and congregants give generously to them. Conservaton is quiet after 10:00 pm. Actually, it is quiet pretty much all the time except for one Saturday night a month. That night, the racetrack on the edge of town attracts some of the fastest stock cars in the region along with over 1,000 loyal fans. The town takes pride in its high school football team, a perennial state championship contender that shares the field on Friday nights with a renowned, amazingly crisp, John Philip Sousa–playing marching band. The restaurants in town are cozy and familiar—they haven't changed their menus in decades and specialize in American food and lots of it. People dress predictably and nicely. Conservatonians are a bit cliquey; they don't take to outsiders much and are especially wary of the residents of the only other town of consequence in the county: Liberalville.

Though Conservatonians would never believe it, Liberalville is a perfect place for some people to live. The schools promote experiences rather than rules and their curriculums change with the latest educational fads and experiments. Houses are an architectural hodge-podge and Liberalites emphasize preserving the character of older buildings even if this means forgoing modern amenities. Wood floors get the nod over wall-to-wall carpeting. Lawns are unlikely to be showered with the copious amounts of chemicals and water needed to maintain thick carpets of green grass. Some residents don't even bother mowing—they just let nature take its course and enjoy the results. The town is light on churches, but has some pretty hip bars and pubs. It also has a community theatre and coffee shops that sponsor interpretive readings and poetry slams. Along with the latest blockbusters, the movie theater makes an effort to bring in award-winning documentaries and foreign films. New restaurants are constantly popping up, and Liberalites can go out for Thai, Ethiopian, Greek, and sushi. The high school's sports teams are a joke. The most successful is the girls' soccer team, and even they only occasionally manage a .500 season. The marching band is equally bad, but the improvisational jazz group is nationally known and regularly wins awards. Local kids in Liberalville are always forming and reforming garage bands, some of which turn out to be very good. Liberalville is never quiet. People come and go at all hours and something is always happening. The loudness extends to fashion; Conservatonians wouldn't be caught dead wearing the togs that Liberalites delight in sporting. Liberalites tend to travel a good deal, sometimes even going abroad. The population of Liberalville is much more diverse than that of Conservaton and it is not uncommon to hear languages other than English being spoken. Liberalites like this and are always interested when new and different people with new and different ways of living move in. In fact, the people of Liberalville are pretty much open to all kinds and all lifestyles with one important exception: Conservatonians.

Conservaton and Liberalville together make up a little less than 50 percent of the county's population. Approximately halfway between the two towns is Middlesboro, which to locals is just the "Middle." Though it covers a large area and includes a big chunk of the county population, the Middle is unincorporated and the people living there agree on little. Indeed, about the only thing that unites the people of the Middle is distaste for Conservaton and Liberalville. Outside of these three population centers, people are scattered widely across the county, most of them living far off the main highway that runs from Conservaton through Middlesboro to Liberalville. People in these outlying areas are a mixed bunch. Some of them take Liberalite or Conservatonian traits to an extreme degree, some (like "libertarians") mix and match these traits, others simply couldn't care less, and still others defy categorization.

The stark differences between Conservaton and Liberalville would fuel little more than a healthy town rivalry except for one thing: Residents of the entire county need to make collective decisions about a range of important public policy matters. Consistent with their starkly contrasting lifestyles and tastes, residents of the two towns display distinct preferences on these policy matters. Conservatonians want to stop migration into the county, to come down hard on county scofflaws, to prohibit gay marriage and gay adoption, to lower payments to the unemployed, to declare English the official and only language of the county, and to require students in all schools to recite the Pledge of Allegiance. Liberalites resist all of these initiatives. They believe criminals should be rehabilitated, not punished; that immigrants should be welcomed and allowed to speak whatever language they want; and that students should decide for themselves whether they want to be allegiant to their county, or anything else for that matter. Pretty much on any and every issue, Liberalites and Conservatonians find themselves on opposite sides of an often-heated argument.

The hardened stylistic and policy differences of the two towns results in a county decision-making process that is polarized and sclerotic. Liberalites and Conservatonians pay to put competing billboards around the county and back-and-forth name-calling is common. While acrimony between the two towns is palpable, those who live in the Middle find the whole schism puzzling, irritating, and immature. More than anything they want the bickering between Conservaton and Liberalville to cease defining life and politics in the county. It's a vain hope—even though the two towns do not account for a majority of the county's population, they do account for a majority of the strong believers. Residents of the county's hinterlands exacerbate the problem; though many loudly decry Liberalville and Conservaton, they often display traits and hold positions similar to one town or the other. Others rail at Liberalites for not being Liberalite enough or excoriate Conservatonians for abandoning the true Conservaton ethos. And so it goes. The Liberalville-Conservaton divide dominates the county's politics just as their less metaphorical progenitors have dominated politics in societies from time immemorial.

Neither town "gets" the values and politics of the other town. They both know in a deep, fundamental, and unshakeable way that their own values and ideas are clearly and obviously the best hope for a better county. Given this, they are puzzled that the denizens of the other town can bear to live there and are shocked whenever anyone expresses a desire to move to that locale. Mistrust between the towns runs deep: Each town believes the other engages in trickery, misleading issue framing, media shenanigans, and forced socialization of youth. Both believe some form of brainwashing must account for the weird, almost cult-like attitudes and behaviors that ripple out from the power centers of the opposing town. How else to explain the propagation of obviously faulty preferences for society and politics?

Residents of each town are convinced the other's propaganda machine must be countered. If people can only be pried away from the lies and presented with the truth, they will reject the despicable ways of the offending town. Cultural misinformation provided at schools, on billboards and television, during dinner and work, and over the Internet thus needs to be corrected so that the truth can be revealed, allowing the division between the two towns to vanish in a sea of equanimity. The residents of each town spend their time alternatively in puzzled disbelief, in sneering contempt, or in quixotic efforts to show residents of the other town the error of their ways. Both sides are convinced that if only people in the other town would confront the facts and analyze them rationally, they would all move.

There is no way that the citizens of Conservaton and Liberalville will ever live happily ever after in political harmony and agreement. Anyone who says different is just naïve. Liberalites and Conservatonians are miffed at the very existence of the other town and believe that, with enough effort, those in the other town can be talked out of their misperceptions and flawed behaviors. The truth, though, is that this approach will never solve the underlying conflict, because the beliefs and behaviors of the other town are being driven by predispositions and not a lack of information. The unfortunate fact that no one in Conservaton or Liberalville recognizes is that predispositions, most definitely including their own, are often more powerful than information, even if that information is truth.

The Earth Is Round; No, It Isn't

One of the most momentous issues in the United States during the last quarter century was how to respond to the September 11, 2001, terrorist attacks on New York and Washington. One major response was the near-immediate invasion of Afghanistan, a nation that, under the leadership of Mullah Omar, gave aid and sanctuary to al-Qaeda mastermind Osama bin Laden. The general attitude seemed to be that if you harbor and protect terrorists who commit mass murder in the United States then you richly deserve a visit from the 82nd Airborne. More controversial was the decision 18 months later to invade Iraq. The warrant for doing so, as enunciated by George W. Bush and his administration was, to say the least, less clear than the justification for invading Afghanistan. Admittedly, Iraqi leader Saddam Hussein was a bad guy. He and his sons were responsible for an untold number of atrocities and brutalities. Yet this fact hardly distinguished him from several other world leaders with blood-soaked resumes whose countries were not invaded. Why go after Hussein in particular? The United States already had one war on its plate and Iraq, by all accounts, had no role in 9/11. Yet the events of 9/11 bred a new, muscular foreign policy that disliked nation states believed to be antithetical to the American way of life. The Bush administration's primary justification for invading Iraq was that Hussein had weapons of mass destruction (WMDs) and that these could end up in the wrong hands. Bush advisor Paul Wolfowitz, for example, said WMDs were the "core reason" for the war with Iraq.

Yet as the Iraq War continued, prewar doubts about the existence of Iraqi WMDs grew. It soon became clear that the teams scouring postinvasion Iraq

could not find WMDs because there weren't any to be found. A key CIA infor-
mant in Iraq admitted that he had lied about the existence of WMDs and then
watched in horror as that lie was used as a justification for the invasion. President Bush soon acknowledged that the claim that Hussein had WMDs was the
product of an "intelligence failure." By 2005, the absence of WMDs at the time
of the U.S. invasion had become an established fact. President Bush admitted
there were none and so did Secretary of Defense Donald Rumsfeld. It was no
longer open for interpretation; Iraq did not have WMDs.

Perhaps there the matter would have remained except for the work of political scientists Robert Shapiro and Yaeli Bloch-Elkon.[1] Using survey data from
2006, they found that many Americans still believed that "Iraq had WMDs at
the time of the U.S. invasion" and the breakdown of believers by party revealed
striking differences. Only 7 percent of self-identified Democrats agreed that
Iraq had WMDs compared to nearly half (45 percent) of Republicans. More
than a quarter (28 percent) of Republicans were convinced that the United
States had actually found WMDs in Iraq. Note that this was not an opinion poll.
People were not being asked whether invading Iraq was a good idea or if Vice
President Dick Cheney was a good guy. They were essentially asked: Iraq had
WMDs at the time of the U.S. invasion, true or false? Responses to this question
can be objectively graded, and a big chunk of people flunked—they replaced
fact with an apparently more ideologically comforting fiction.

Conservatives are certainly not the only ones who have trouble with the
facts. Presented with evidence of the Soviet army's invasion of Hungary in 1956
and Czechoslovakia in 1968, members of the French Communist party often
denied such events occurred. Note that the French Communist party was not a
fringe group in the 1950s and 1960s; in fact, during those years, it was the largest
left-of-center party in French politics and claimed the political loyalty of a big
chunk of the electorate. It was a mainstream party whose ideology and values
made it averse to acknowledging any flaws in the workers' paradise that was the
Soviet Union. So when it came to Soviet strong-arm tactics in East Europe they
would simply claim the information—that is, the clearly established fact—was a
fiction.[2] As researchers presented more and more evidence, including pictures
of Soviet troops beating up and sometimes killing Czech workers, the French
Communists, often with great anguish, would sometimes reverse course and
very reluctantly acknowledge the truth. For some, the evidence concerning

Soviet actions was so directly contrary to their worldview that they became physically ill. Opting for comforting fiction over unassailable fact is clearly not limited to one end of the ideological spectrum.

That people are often misinformed about politics is hardly news. Entire books have been written on how people in general and Americans in particular are factually challenged when it comes to politics.[3] One oft-noted error concerns the percent of U.S. government expenditures going to foreign aid. The actual figure is vanishingly small, well under 1 percent, yet survey respondents consistently put it much higher, sometimes well into double digits. Respondents to one survey estimated that an astonishing one of every four dollars spent by the federal government went to foreign aid. Then they said their preference would be to "cut" that figure to about 10 percent. This would take some Alice in Wonderland math; essentially they were asking that foreign aid spending be cut by increasing it 1500 percent.[4] Our favorite factual error comes from a survey asking people to identify the source of the quote, "[F]rom each according to his ability; to each according to his need." Forty-five percent of Americans proudly assert that this phrase is the U.S. Constitution when it was actually written by Karl Marx, who no doubt would take some glee in this particular mistake.[5]

Still, general ignorance is not what is interesting about the Shapiro and Bloch-Elkon research because their study is not really about how little people know about politics. What they demonstrate is that in politics, ignorance is not random; factual errors are targeted in a particular direction. Conservatives rewrite history to justify the decision of a Republican administration to enter into a war that, by the reasoning of its architects, was quite possibly unnecessary. Communists rewrite the history of the Prague Spring to expunge the murderous culpability of their model state: the Soviet Union. If the facts get in the way of your preferred worldview, just unwittingly "misremember" the facts. This pattern of behavior is consistent with recent research showing that once people adopt a preferred political candidate, new negative information about that candidate leads them to intensify rather than lessen their support.[6]

It gets worse. In the largest study ever of "false memory," scholars presented volunteers with accounts of five unrelated news events, each accompanied by a photograph. Unbeknownst to research participants, one was a complete fabrication with a photoshopped accompanying image. Yet half of the people said they distinctly remembered the fake event happening; 27 percent even said they saw the nonexistent event on the news.[7] The pattern of these false memories was not random. One of the fake events was President Obama shaking hands with Iranian President Mahmoud Ahmadinejad—Holocaust denier, main proponent of the Iranian nuclear program, and sworn enemy of Israel. Another concerned President George W. Bush vacationing with a famous athlete during the midst of Hurricane Katrina. Both were completely fake. President Bush was in the White House when Katrina ravaged the Gulf Coast and President Obama has never joined hands with Ahmadinejad. Yet many survey participants not only confidently remembered those two events, they explained just how they felt at the time they first learned of the event. The punchline, of course, is that conservatives were much more likely than liberals to falsely remember the Obama-Ahmadinejad handshake (by better than a 2 to 1 ratio), while liberals were significantly more likely than conservatives to falsely remember the inappropriately timed Bush vacation. So not only do people refuse to remember unflattering things about those with shared predispositions, they also make up unflattering things about those who have opposing political predispositions. People with different politically relevant predispositions appear to live in worlds with distinct sets of facts. It would seem Moynihan's famous statement quoted at the beginning of this chapter does not always apply.

Driven by their fundamental differences in predispositions, liberals and conservatives believe the facts that support their predispositions even when they are not real facts, and once people have erroneous beliefs it is extremely difficult to correct them, since their instinct is often to dig in their heels. This being the case, the logical inference is that it will be virtually impossible to get liberals to become conservatives or conservatives to become liberals. Is this true? Do we choose our political beliefs or do they emerge out of predispositions that are at best only partially under our control?

Sex and Politics: Do We Have a Choice?

Actress Cynthia Nixon, better known to millions as *Sex and the City's* Miranda Hobbes, says that she is gay by choice. "Why can't it be a choice?" she asks.[8] Well, clearly for her it can. She had a long-time relationship with Danny Mozes, who fathered two of her children, and she did not switch teams, as it were, until well into adulthood. Fair enough. The United States is a free country and she exercised that freedom to opt into a same-sex relationship. Making that choice, though, has irked many people. Liberal gay rights advocates worry that Nixon is willingly falling into a "right-wing trap." John Arovosis, a Democratic political consultant and gay activist, argues, "When the religious right says it's a choice, they mean you quite literally choose your sexual orientation, you can change it at will and that's bull."[9] Liberals like Arovosis seem quite comfortable with sexual orientation being biologically based rather than a lifestyle choice reflecting cultural constraints and our mood of the day, but they are often decidedly uncomfortable with any other orientation being biologically based.

To see what we mean, let's take a short walk back in time. Forty years ago, Harvard biologist E. O. Wilson argued that social traits like cooperation were almost certainly shaped by evolution.[10] As long as Wilson was talking about ants, his ideas were given a respectful reception, which was not surprising given that they made theoretical sense and were backed by solid evidence. But when he extended the same ideas to humans, in effect saying that human nature was rooted in biology, he triggered a huge backlash from the political left. Well-known liberal intellectuals like Stephen Jay Gould and Richard Lewontin lined up to bash the notion that social characteristics were biologically rooted. They believed Wilson was guilty of "biological determinism," or at least social Darwinism. The International Committee against Racism (CAR) claimed that by encouraging "biological and genetic explanations for racism, war and genocide," Wilson "exonerates and protects the groups and individuals who have carried out and benefited from these crimes."[11] Left-leaning academic groups like Science for the People denounced the notion of a meaningful genetic basis for social traits, and scholars from various disciplines banded together to write letters to national publications declaring that "(sociobiology) has no scientific support, and . . . upholds the concept of a world with social arrangements remarkably similar to the world which E. O. Wilson inhabits."[12] In a nutshell, Wilson's opponents accused him of using science to justify the social status quo and anyone who does that must be a conservative—except Wilson isn't.[13] Things got so bad that at one conference a group of CAR activists interrupted a presentation by Wilson, called him a racist, and dumped a pitcher of water over his head.[14]

The same battle lines cropped up two decades later when right-leaning intellectuals Richard Herrnstein and Charles Murray (a Harvard psychologist and a think-tank political scientist) caused a huge stink by claiming that intelligence was genetically influenced and that high-IQ types were becoming a distinct social group that they referred to as the "cognitive elite."[15] What really got Herrnstein and Murray in hot water was their argument that those who were not in the cognitive elite tended to fall into certain sociodemographic groups—for example, they were more likely to be criminals and quite a bit less likely to be well-off or white. Gould and a long list of liberal luminaries once again mounted the barricades to defend the notion that human nature is a pure product of social environment and has no heritable or biological basis.[16]

This war over the biological basis of social traits goes on today. Rather than rehash the pointless dispute over nature and nurture, we want to highlight the hypocrisy of both sides. Liberals fulminate that researchers are mistaking right-wing ideology for science when they find that intelligence is genetically influenced, but apparently the science is high quality when it suggests that sexual preference is genetically influenced. Conservatives argue that people need to brace up and face the implications of scientific research on the heritability of intelligence but their face-the-data stoicism goes into reverse when it comes to studies suggesting sexual orientation is heritable.

So both sides agree that socially relevant traits are biologically influenced—they just disagree about which ones. In reality, of course, the facts of biology are not structured to please one political side or the other. A wide swath of socially relevant traits—sexual behavior, intelligence, personality, and a whole lot more—appears to be influenced by both nature *and* nurture, not one or the other. This idea is steadily becoming conventional wisdom regarding a growing number of social behaviors, but there is one big exception: politics.

Politics is a last redoubt for hard-core supporters of a version of human exceptionalism that maintains biology always applies to other species but not always to super-special homo sapiens. And one area in which it certainly does not apply, this argument goes, is personal political temperaments. The belief is that politics is a purely cultural construct and is therefore immune from biology. Politics is the Alamo for people who deny the relevance of biology and it must be defended against assaults from Santa Annas like us, who are coming over the walls in increasing numbers, waving EMG sensors, asking for saliva samples, and showing people pictures of poo to see if it makes them sweat.

As far as we can tell, the fierce determination of the defenders on the wall is motivated at least in part by the understanding that ideas have consequences. The sad and often sickening history of the application of biology and evolution to human affairs gives legitimate cause for concern; racism, sexism, classism, and assorted other odious isms have all been supported by the antecedents of sociobiology. The eugenics movement of the first half of the twentieth century, for example, was backed by big scientific names who assured us that it was a useful, even necessary, application of new biological knowledge to human affairs. Karl Pearson, who developed the statistical methods of correlation we introduced in Chapter 1, was a respected mainstream public intellectual in the early part of the twentieth century. He was also a big fan of social Darwinism and an enthusiastic eugenicist; his view of the ideal country was one with a population "recruited from the better stocks" so that it could keep its competitive edge "by way of war with inferior races."[17]

Of course, liberals sometimes like to forget that their favorite opposing "big idea"—that people are shaped exclusively by their environment—has led to its own share of atrocities. To return to the example mentioned in Chapter 3, Mao's notion that people could be changed just by forcing them to move from fetid cities to noble pastures resulted in millions of deaths. If the conclusion is that ideas matter, we are on board; if it is that giving weight to environmental influences on behavior is good but doing the same for biological influences is bad, we are disembarking. An unrelenting faith in the ability of social context to mold behavior is hardly a source of tolerance, as gay rights advocates know all too well. All knowledge can be put to good or bad uses. The knowledge (in the form of empirical findings) presented in this book is neither inherently dangerous nor universally depressing. Let us try to convince you that acknowledging the role of biology and deep psychology in shaping political orientations could, under the right circumstances, do some good.

Taking Political Predispositions Seriously

The material presented in this book cannot eliminate the forces behind diverse political orientations and will not end political polarization or pave the way for a universally productive and trusted political system. Nonetheless, it has the potential to improve the political climate. This book has been about political predispositions. These predispositions, as we hope is clear by now, run much deeper than the sources of attitudes presumed by the citizens of Conservaton and Liberalville to shape beliefs: billboards, schools, families, and talking heads.

As depicted in the top part of Figure 9.1, because of genetics, early development, and subsequent life experiences, people carry with them distinct brain response patterns, sympathetic nervous systems, values, moral foundations, negativity biases, strategies of information search, tastes, preferences, and, it would appear, even different "facts." These differences coalesce into what we have been calling predispositions or, to be more precise, behaviorally relevant biological predispositions (BRBPs for short).[18] Person-to-person variations in predispositions connect directly to variations in "neuroception," a term coined by psychophysiologists to describe the fact that physiologies are constantly scanning and evaluating the environment and generating signals about what is safe and approachable and what is threatening and deserving of a wide berth.[19]

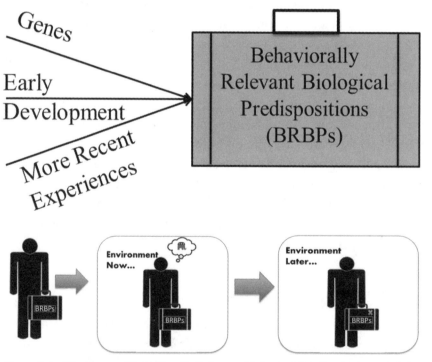

Figure 9.1   The Bases of Behaviorally Relevant Biological Predispositions and the Interactions with the Environment

Neuroception often takes place outside of conscious awareness but makes itself known in our emotions, in whether we feel that something is worth a taste, revolting, the right thing to do, or just not right. In other words, neuroception is the way people perceive and experience the world.

Individual differences in biology connect to individual differences in neuroception, and since neuroception monitors the sociopolitical environment, those biological differences can and do correlate with political attitudes. In other words, the structure, wiring, and processing of their central and autonomic nervous systems leads some people to find certain stimuli instinctively and intuitively appealing while others find them repellent, just as it leads certain people to care about a given outcome while others do not care at all. Since public policies can alter the likelihood that particular stimuli will make an appearance, it is not surprising that these physical characteristics affect politics. As a result, in a given situation, preset response tendencies (that is, the predispositions we carry with us) shape political views and opinions, even though people often deny it.[20] Once these predispositions exist, people try to mold the world to fit them (rather than the other way around) and this is why the two ideological camps often end up with different sets of facts. Liberals mold the world their way, conservatives mold it theirs, and the result is a big divide on many different topics.

Most people seem to wander around thinking that their political views are "common sense" or "normal" and that a majority of their fellow homo sapiens either already agree with them or would agree if they rationally thought things through. Psychologists term this "false consensus." Moreover, people assume that those who do subscribe to aberrant political viewpoints could be persuaded otherwise with only a handful of simple ingredients: a few facts, a dollop of logic, and a pinch of persuasion. Mix that together and feed it to reasonable people, and they will adopt your political perspective, whatever that happens to be. Anyone who persists in holding the opposite viewpoint must be irrational or pigheaded, so no wonder their political ideas are half-baked. This set of beliefs about the nature of political differences is why Liberalites and Conservatons spend so much time gridlocked and yelling at each other. They fundamentally misunderstand why the other side holds the views it does.

A central implication of the evidence we have summarized is that people are always going to have different points of view and not merely because they are information deficient or obdurate. As such, our message is perfectly consistent with that offered by others,[21] but takes the point to a deeper level by tracing the differences in such concepts as moral foundations and personal values all the way to sub-threshold physiological and cognitive traits and biases (and possibly even genetics) of which people have no conscious awareness. Even so, this does not mean change is impossible.

As indicated in the bottom half of Figure 9.1, when certain environments act on an individual's behaviorally relevant biological predispositions, particular thoughts, decisions, or actions result. Maybe the person demonstrates an affinity for one political party or the other. This earlier environmental context as well as the resultant thought, decision, or action it engendered might not have any lasting effect on the person's BRBPs and therefore would not be expected to lead to any alteration in subsequent thoughts, decisions, and actions. (In fact, we would argue that this is the norm and that is why political views are so stubbornly held and rationality and compromise are so difficult to locate.) However, on other occasions, the BRBPs might be altered (indicated by the "X" on the suitcase in the figure) by a poignant and powerful feature of the environment—perhaps not altered by much, but enough that slightly modified predispositions will be carried into later environments. Predispositions can change, but they do not do so often.

The major reason to retain and even refine your powers of political persuasion is not the likelihood that you can convert the politically predisposed (though being able to put yourself in their shoes might help some), but the possibility of influencing the large number of people who lack such predispositions. They might actually respond to good arguments and fresh evidence. We recommend not wasting your breath on those who are predisposed toward political positions that run the opposite of yours. The payoffs of working on this group are just too small. You may derive some twisted satisfaction from trying to move the unmovable, but the effort tends to pollute the whole political climate. Slices of the population on both the political left and the political right are predisposed, and therefore for all intents and purposes unpersuadable. Unfortunately, those who have these predispositions tend to be the ones who are the most politically motivated, the upshot being that the most intransigent among us tend to have disproportionate influence on the nature of the political system.

On its face, the existence of biologically grounded political predisposi-tions seems an incredibly depressing situation for those wanting to improve political arrangements in the United States or elsewhere. However, we believe a silver lining accompanies the increasingly documented existence of political predispositions. Though political predispositions distort facts, hinder political communication, sow mistrust, and even initiate violence, they bring benefits as well—as long as we are made aware of their existence. Properly handled, the realization that political views are shaped by predispositions can help each of us to understand ourselves, understand our political opponents, and understand the best design for political systems. If acknowledging the fact that political temperament is traceable in part to biology requires modifications of both canonical thinking in social science disciplines and hackneyed interpretations of human history, as Professor Charney suggests in the quote that opens this chapter, we say that isn't all bad. After all, look where the old interpretations of history and applications of the social sciences got us. It is high time to embrace the real version of humanity, not the one that makes us feel good but in actually is quite unhealthy in the long run.

Know Thyself; Know Thy Enemy

If you are a conservative, do not read the next few paragraphs, as they are designed solely for liberals and are meant as a private counseling session intended for them only. Liberals: Quit wasting your time spluttering about the ignorance of con-servatives or trying to convert any and all of them. As F. Scott Fitzgerald might have put it, "The very conservative are different from you and me." Where you see a titillating curiosity, they see an imminent danger; where you see something potentially edible (with the right mole), they see disgustingly spoiled produce; where you see an excuse to hire a domestic worker, they see unmitigated chaos; where you see intriguing ambiguity, they see debilitating uncertainty. They spend more time than you focusing on negative events—particularly negativ-ity that is tangible and immediate. They see problems that are not there. They "remember" events and visions that never were. They refrain from seeking new information simply because it might not be information that is helpful or con-firming. They are comfortable with revered and long-established sources of authority such as religious orthodoxy and the words of the country's founders.

On the other hand, anything that reeks of human discretion, like modern governments and a broad application of scientific investigation, is suspect. Their first instinct is to assume those in faraway lands have questionable values, do not share our country's interests and goals, and should not be trusted. Conservatives prefer established ways of doing things and have less craving for new experiences—culinary, social, literary, artistic, and travel—than you do.

Their enhanced focus on negative events and situations should not be mistaken for fear. Au contraire! They do not run from the negative. They attend to it, eye it warily, and ponder how best to minimize its influence and impact. They don't like being told what to do, especially by people who are not part of their in-group, because they don't trust the judgment of other human beings. They think the only hope for mankind is to embed it in hierarchies and rules, to remove individuality and discretion by following inviolate texts and the dictates of the free market that, thanks to Adam Smith's invisible hand, work automatically on the basis of supply and demand. They think rules are good as long as they derive from the proper authorities.

You should not expect them to change, but rather should work with who they are. Try to see the world from their perspective. Work at thinking like conservatives think and experiencing what conservatives experience. Enter their world not by actually going undercover but by attempting to adopt the psychological mindsets that make conservatives conservative. If that is not doing it for you, come to our lab and, for a small fee, we will condition you to attend like a conservative to negative stimuli, looming disorder, and mild ambiguity. You will know you have succeeded when you "dream conservative."

This is where conservatives need to come back and liberals need to leave. More specifically, if you are a liberal, do not read the next few paragraphs as they are designed solely for conservatives and are meant as a private counseling session for them only. Conservatives: Quit wasting your time spluttering about the ignorance of liberals or trying to convert any and all of them. To paraphrase Fitzgerald, "The very liberal are different from you and me." Where you see an imminent danger, they see a titillating curiosity; where you see disgusting spoiled produce, they see something potentially edible; where you see unmitigated chaos, they see an excuse to hire a domestic worker; where you see debilitating uncertainty, they see intriguing ambiguity. They don't pay nearly as much attention as you do to negative situations and potentialities and, if they do worry at all about the negative, they seem strangely unmoved by the immediate threat of malevolent human beings. Sometimes it seems as though they worry more about climate change and endangered species than terrorism and crime. They are firmly convinced that, despite all evidence to the contrary, humans can change under the right circumstances.

All this makes liberals far more trusting than they have any right to be, but it is important to realize that this is not because they are foolish or lazy but rather because they are structured in such a way that prevents them from appreciating the obvious dangers swirling about. They seek out new information even without knowing where it might lead and even when that new information might be contradicted by even newer information. None of this particularly bothers them, as they just like the idea of moving from new thing to new thing as though novelty were its own reward. They really believe that government programs and the like will change things for the better and they are suspicious of the tried and true. They are convinced that the traditional approaches created big problems, problems that are remediable by embracing the untried and new. Their first instinct is to assume individuals in faraway lands are trustworthy. Hierarchies, on the other hand, such as those typifying the military, organized religion, and corporations, are objects of their suspicions. They love experiences that might take them off the beaten track. They seem not to look before they leap.

Their eagerness to try new approaches and experiences should not be mistaken for reckless hedonism. On the contrary! Liberals spend a good deal of their time trying to understand other people, even worrying about them. The circumference of their circle of concern extends around the globe and even incorporates nonhuman life forms. They don't seem to consider, let alone mind, the fact that this openness raises the possibility that they could be taken in by evildoers. Because they think the human condition is perfectible, they are always trying new approaches, which usually fail. But this fact seems not to dissuade liberals from turning right around and trying something else. They like to be surprised by their food, their literature, their art, and the places they visit.

Liberals "just don't get it" and you should not expect this to change because for liberals there is nothing to "get." Quit wasting your time explaining to them the dangers of rampant immigration, overseas threats, and moral decay. Nothing you say will lead them to take these matters as seriously as you do. Rather, take what you now know about them and work with it. Try to see the world the way they see it. Hold in abeyance your knowledge that threats are real and try not to be bothered by what will initially feel to you as vulnerability and carelessness. Practice not fixating on the negative and work at enjoying new and unexpected experiences. Do this not with the intention of becoming a liberal but with the intention of better understanding them. Work at thinking like liberals think and experiencing what liberals experience. Enter their world not by actually going undercover but by attempting to adopt the psychological mindsets that make liberals liberal. If that is not doing it for you, come to our lab and, for a small fee, we will condition you to attend more than you currently do to positive stimuli rather than threats, looming disorder, and nagging ambiguity. You will know you have succeeded when you "dream liberal."

A Zebra Can't Change Its Stripes

Okay, reading in rounds is done and we hope everyone is back. Actually, we know you cheated, and we are glad you did because the goal here is to help our readers more deeply understand differences, and especially political differences. Before you accuse us of painting with too broad a brush, don't forget to think probabilistically. Obviously, not all conservatives and not all liberals fit the descriptions above (thank goodness!), but the general tendencies appear over and over. The larger point is that those with predispositions counter to yours do not see what you see, fear what you fear, love what you love, smell what you smell, remember what you remember, taste what you taste, want what you want, or think how you think. These differences run so deep that they are biologically grounded and, as such, cannot be changed quickly. Since political beliefs flow out of these predispositions, this means that they, too, cannot be changed quickly. It is our conviction that making an effort to understand the nature and depth of political mindsets will be beneficial since it is always good to better appreciate those with whom we are sharing the planet. Just as learning a second language assists in coming to grips with your native tongue by putting aspects of language in perspective, learning a second political orientation also puts your native orientation in perspective and deepens understanding.

In addition to self-improvement, taking predispositions seriously can improve understanding of others and therefore can enhance the state of political discourse. Recognizing that the maddeningly incorrect views of your political opponents are due less to their unencumbered choices than to traits they have little choice but to endure cannot help but increase tolerance and acceptance. Think of the improvements resulting from the recognition that being left-handed is not a choice resulting from flawed character but instead is the product of a biological (in this case heritable) disposition. Teachers are no longer disrupting classrooms and wasting time (not to mention demeaning 12 percent of the student body) by trying to beat the left-handedness out of left-handers. The entire learning environment has improved as a result. We look forward to the day when liberals are not trying to beat the conservative out of conservatives and conservatives are not trying to beat the liberal out of liberals, as we believe parallel improvements in the political system will be in evidence.

A more commonly invoked illustration of the good that can come by acknowledging a role for genetics and biology in human social behavior is sexual orientation. Gallup periodically polls people on whether they think homosexuality is a product of nature or nurture, something that a person is born with or the result of something in their environment. In 1977, only 13 percent of people believed being gay was innate, while 56 percent attributed it to upbringing. In 2012, 40 percent believed sexual orientation was something you were born with and only 35 percent attributed it to upbringing. The different beliefs in the causes of homosexuality have fairly stark implications for attitudes on gay rights. Roughly two thirds of people who think being gay is a product of upbringing and the broader environment say that homosexuality is not an acceptable lifestyle. On the other hand, more than three quarters of the people who believe homosexuality has a biological basis believe it is an acceptable life-style.[22] Figure 9.2 tracks Gallup data over time and shows that as more people believe homosexuality is something you are born with, the percent believing homosexuality is an acceptable alternative lifestyle also goes up.

Given those numbers you get a sense of why gay rights activists get alarmed when the Cynthia Nixons of the world announce their sexual preference is purely a personal choice. The bottom line is that people who believe sexual orientation is biologically based are much more likely to be accepting of gay rights.[23] Americans became more accepting of gay lifestyles and gay rights because they started to accept the growing evidence that sexual orientation is less a moral choice than simply a part of people's biology.[24] This shift in perceptions of the

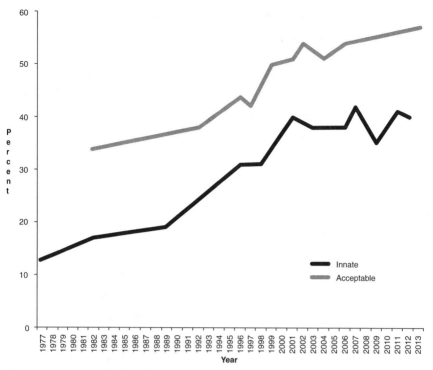

Figure 9.2   Percent of People Believing Homosexuality Is Innate and Homosexuality Is an Acceptable Alternative Lifestyle

source of homosexuality almost certainly helped the political conversation over gay rights to take place within the appropriate democratic and legal context. This is a stark example of an awareness of the role of biologically based predispositions helping to generate tolerance for different points of view.

In a roughly analogous fashion, this is the way people should be thinking about the sources of differing political orientations, and when they do, we predict that the implications will be similar. If recognizing that sexual orientation is anchored partially in biology leads to greater tolerance of different lifestyle choices, recognizing that political ideology is also tied to biology will lead to greater tolerance of different political viewpoints. We don't mean tolerance in an everybody-is-special PC sort of way; we mean tolerance in the sense of an acceptance that the world is always going to have people with political temperaments very different from our own. A defiant chant of the gay rights movement is, "We're here. We're queer. Get used to it!" It is the same for ideology. There will always be people with different political orientations from yours, people whose viewpoints you revile. Get used to it, and for the sake of all of us, go one step further—accept it.

This kind of acceptance directed at predispositionally driven variations in political beliefs would not mean you have become a traitor to the cause. We need to get past the stage where liberals/conservatives are in a contest to show that they are the most outraged by their ideological opponents. It would not even mean that you were any less convinced that your political opponents are wrong. You would just be acknowledging that the reason they are wrong is largely beyond their control. This in itself is a major step forward. Accept that the main reason your political opponents hold the views they do is not laziness, a lack of information, or willfully bad judgment, but rather physiological and psychological contours that are fundamentally different from yours. If you had the same predispositions they do, it is likely you would have political opinions similar to theirs. Whenever you meet a conservative/liberal your response should not be, "What a shallow idiot," but "There but for the grace of God go I."

In his book *The Righteous Mind*, Jonathan Haidt describes making just the sort of adaption we are suggesting. Haidt discusses an extended stay in India, a socially conservative society with many traits (patriarchy, hierarchy, extreme inequality) at odds with his beliefs. Though initially disturbed, Haidt also recognized and appreciated the courtesy and hospitality of his hosts and the Indian people more generally. He began to see the dense "moral matrix" and how it supported a vibrant, family-centered social order. He returned to America to find, somewhat to his own surprise, that he was no longer reflexively puzzled by or angry at social conservatives. He still didn't agree with their policy stands, but found himself less viscerally committed "to reaching the conclusion that righteous anger demands: we are right, they are wrong."[25] What we are suggesting is that you don't need an extended stay in a foreign culture to get to the same point. Just recognize the existence of biologically based predispositions and the implications this recognition holds for politics, and you are most of the way there.

So believe in your opinions; they are likely the correct opinions for you. But recognize that they are not the correct ones for everybody. Be humble about them and recognize that they will not and cannot lead to the kind of society everyone wants because not everyone has the same perceptions of reality and therefore of the most desirable social arrangements. If everyone saw and experienced the same world, you could be more confident that your beliefs were correct and should be broadly applied. Since this is not the case, your beliefs are not such a big deal and humility should be the order of the realm. If people internalized these facts, political debate would be different and better.

A deep irony in all this is that, by demonstrating the major differences in people's visions and experiences of the world, science is increasingly supporting a concept long favored by rabidly antiscience deconstructionists. Deconstructionists believe there is no objective reality because everything is in the eye of the beholder; that there is no text until the reader brings his or her own interpretations and experiences to it. Our take on this improbable confluence is likely to alienate both sides. The deconstructionists are right in positing that, because of the significant variations in neuroception, when it comes to topics such as politics there often seems to be no objective reality. However, these differences across people can (and should) be studied scientifically in order to understand the nature and implications of variations in people's perceptions. Come to think of it, that is exactly what we are doing.

Talking Conservative/Speaking Liberal

Historically, the two sides have a hard time talking to each other because they often speak a different language. This sort of argument has been made most famously by George Lakoff, a cognitive linguist at the University of California–Berkeley. Conservatives, argues Lakoff, frame their arguments in the language of the "strict father," a metaphor for their preferred relationship between government and the governed. Liberals, on the other hand, use the language of the "nurturing parent." Lakoff contends that their different languages make it difficult for conservatives and liberals to comprehend each other. He also argues that the conservative strict father approach resonates more with the broad electorate, which is why conservatives (at the time he was writing) were doing a better job capturing the support of middle-of-the-road voters. Lakoff suggests liberals take a cue from conservatives and adopt the language of the strict father.[26]

Though our predispositions perspective is consistent with Lakoff's notion that conservatives and liberals see, understand, and especially describe the world differently, its tactical implications are quite different. We doubt the effectiveness of liberals speaking a language other than their native tongue. Liberals attempting to talk tough and strict, even though doing so defies their predispositions, are likely to appear as inauthentic as Michael Dukakis in a tank or Barack Obama shooting skeet. Moreover, even if liberal politicians attract a few moderates and conservatives by becoming strict fathers and mothers, liberal followers are even more likely to turn off and tune out. Likewise, conservatives suddenly trying to come off as nurturing would be no more effective than Dick Cheney attempting to smile or 2012 Republican vice presidential nominee Paul Ryan's widely ridiculed soup kitchen photo op. Predispositions are real, and ordinary people are amazingly good at intuiting who does and does not share their predispositions, even though they may not be conscious that they are making these judgments.

When people feel a candidate is one of them, they will cut that candidate an amazing amount of slack on policy and personal matters. If they sense something is off, support will be half-hearted at best. Consider the strained relationship of 2012 Republican presidential candidate Mitt Romney with true conservatives. It wasn't just his Mormonism or his embrace of healthcare reform when he was governor of Massachusetts; more generally, he just did not seem to have that conservative swagger, vocabulary, mindset, and look in the eye. True-believing conservatives serially fell in behind every other possible potential nominee in the primary race and then backed Mitt Romney only when each of those alternatives imploded and nobody else was left. The only thing that dispositional conservatives found exciting about Governor Romney in the general election was that he was not Barack Obama. Regardless of policy stances, he would never be one of them.

The point we are making is that, unlike Lakoff, we do not believe that predispositions can be "gamed." Artificially adjusting adjectives will not sway those who have predispositions, and the middle ground is a mixed and unpredictable bag. Pretending to be something you are not is rarely a successful strategy in the long run. Besides, though the cycle will no doubt continue its ups and downs, recent electoral results suggest that the "liberal" nurturing lingo may not be as big a loser as Lakoff implies. We do not see the existence of predispositions as a particularly propitious platform for one group to bamboozle the other (or even to bamboozle those largely devoid of predispositions—they are pretty good at spotting fakes too). Understanding predispositions does not necessarily create an opportunity to achieve a strategic political advantage, but rather constitutes a much more mundane but perhaps more important opportunity to improve acceptance. This acceptance is more crucial than ever in our increasingly interconnected world. As Shankar Vedantam puts it, "[O]ur mental quirks and biases once affected only ourselves and those in our vicinity. Today they affect people in distant lands and generations yet unborn. . . .Today, subtle biases in faraway minds produce real storms in our lives."[27]

Building a Better Mousetrap

The predispositions argument does not mean that you need to agree with viewpoints that contrast with your own—far from it—but it does have several potentially valuable implications for the type of political system that might best manage these disparate predispositions. For starters, longing for a political system devoid of ideological and partisan differences is pointless. George Washington never actually delivered his famous farewell address (it was merely published in leading news outlets), but the most remarked passage in it laments "the mischiefs of the spirit of faction," which at the time was taken to mean groups of people united by a common impulse, passion, or interest. Washington noted that factions "distract public councils . . . enfeeble public administration . . . agitate the community with ill-founded jealousies and false alarms . . . [and] kindle animosity of one against another." We concur with Washington—factions do all this and more—but we disagree with his preferred solution, which seems to consist of little more than a plea that factions "be discouraged."

Political factions are built on the foundation of biologically instantiated predispositions. As a result, you can "discourage" all the live-long day if you want, but they will not go away. They will, however, take on contours reflecting the irreconcilable antagonism between the forces of tradition and innovation. Though we share Washington's frustrations with factions—they are strong, resilient, and irritating—we prefer the approach of another founder, James Madison. Madison also was frustrated with factions (not surprising, since he had a big hand in writing Washington's Farewell Address), but in his masterpiece, Federalist #10, he recognized that "the latent causes of faction are sown in the nature of man." Madison is so on target that we can't resist allowing him to continue: "A zeal for different opinions . . . [has rendered people] . . . much more disposed to vex and oppress each other than to cooperate for the common good." And finally, "[S]o strong is this propensity to fall into mutual animosities that where no substantial occasion presents itself, the most frivolous and fanciful distinctions have been sufficient to kindle their unfriendly passions and excite their most violent conflicts."[28] We could not agree more; indeed, Madison could be writing an incisive commentary on modern American campaigns. If there is nothing important about which to argue, something frivolous and fanciful will do just fine!

Madison suggests two mechanisms to do away with factions and bring political combatants like Liberalville and Conservaton into unity and amity. The first option is to destroy the liberty that allows people to pursue their differing impulses and interests; the second is to force every citizen to hold the same opinions and passions. Noting that the first is a cure worse than the disease and the second is wholly "impracticable," Madison quickly proceeds to a discussion of the best ways to mitigate the effects of faction. To cut to the chase, Madison's central suggestion for governing in the face of factions is the implementation of representative rather than direct democracy. He believed that people should not be expected to make political decisions for themselves, but rather urged that a small number of citizens be chosen to represent the differing wishes of the many, so that they could "refine and enlarge" the public views. Madison thought representative democracy was particularly valuable, even essential, to the maintenance of large political systems where direct democracy would be both dangerous and impossible. The basic logic is that since people are inspired to form factions, we must empower a select number of individuals who, if they want to keep their positions, have to pay attention to the interests of more than their own faction. The system Madison created may be failing to accomplish this, as the leaders themselves get pulled ever more into the factional morass above which Madison expected them to rise, but his instinct to mitigate factions rather than wish them away is well-founded

Representative democracy is not guaranteed to solve the problem of factions; direct democracy, though, is guaranteed to exacerbate the problem of factions. Consider a variant of direct democracy known as deliberative democracy, which has attracted extensive attention from political scientists. The basic idea is to get ordinary people together so they can hash through particular questions or issues. In certain iterations, the citizens are given access to information and experts. The hope is that the participants will distill a consensus that accurately approximates the "will of the people"; in other words, a representation of what public opinion would look like if people were informed, engaged, and stripped of false consensus. Various deliberative democracy experiments have been conducted and sometimes report a softening of harsh opinions.[29] Unfortunately, research also shows that any impact of deliberative democratic processes is contingent and limited. It is contingent both on the type of issue under discussion being an easy or "non-zero-sum" issue and on the participants being unusually interested in politics. Even then, any effect is limited to an unrealistically short period of time after the experiment. In truth, most people do not change much even after hearing the thoughts of fellow citizens or being provided with a little more information. Tellingly, the only changes of note tend to come from those who are truly undecided going in.[30]

In short, even when social scientists have given consensus-building the old college try in the most congenial of circumstances, very strong—perhaps irreconcilable—differences in political views remain. We are not surprised, since changing predisposed people's minds, especially on issues relating to the bedrock dilemmas of politics, is not done easily. Other variants of direct democracy, including New England–style town meetings, have similar problems, and the existence of what Washington and Madison call factions and what we call predispositions is a reason to embrace representative government as our best hope. The key is for people to recognize that government is taking place in the context of these widely varying predispositions. When you do not get what you want, it is important to recognize that it is not because of a conspiracy or some structural flaw but because some people have vastly different predispositions than you do.[31]

The predispositions perspective does suggest (or more accurately, reinforce) arguments for institutional reforms that might reduce or at least better manage ideological conflict. These mostly deal with the structural elements of political systems that allow the Conservatons and Liberalites of the world to disproportionately define choices about collective action. Obvious examples in the United States are primary elections and redistricting. Compared to general elections, primary elections tend to be low-profile affairs. Those who cast ballots in these elections, and the groups that recruit and support candidates, tend to be more ideological than the general electorate. This helps explain the lament of many American voters put off by what they see as the overly ideological bias of both the candidates they have to choose between in a general election. The more moderate candidates—the ones more likely to focus on political practicalities than partisan point scoring—often don't make it past the primary. A reformed system that allows residents of the Middle a more meaningful role than choosing between a hard-right Conservaton and a hard-left Liberalite would make more room for candidates and politics that can bridge the gaps.

Similarly, in many states the decennial chore of redrawing the geographical boundaries of congressional districts highlights and exacerbates ideological predispositions rather than helping to mediate between them. Redistricting often amounts to little more than a brutal cartographic war between partisan camps jockeying for electoral advantage; in effect, a duel with maps in which the winner gets to institutionalize the interests of their predispositions. If you want democratic politics to have a more practical tilt to it, then it's probably not a good idea to put the job of geographically defining representation into the hands of people and groups with strong incentives to let partisan self-interests push aside all else.

While understanding the relevance of predispositions supports these sorts of institutional reforms, its real message is aimed at individuals. No magic institutional formula can make divided politics go away. Given the evidence presented in this book, who do you think is going to take the most interest and be the most involved in hammering out the specifics of the institutional reforms we just mentioned? When it comes to mass scale politics, it is impossible to avoid the implications of predispositions and the best that can be done is to manage these predispositions in a way that insures we count, rather than bash, heads to resolve differences. The message of the predispositions argument to those seeking a form of low-conflict politics based on mutual cooperation, interest and goals is this: Grow up. The predispositions argument, if nothing else, explains why democratic politics is so unpalatable—and also so deeply necessary.

Caveat Iterum

Several of the caveats we have been stressing throughout this book deserve repeating one last time. We will mention three. First, remember that though predispositions have a certain timeless quality to them, the issues of the day superimposed on predispositions vary widely. Fifty years ago the issue of interracial marriage was big. When the Supreme Court finally prohibited states from enforcing antimiscegenation laws in 1967, 16 states were affected and only 20 percent of the public approved of interracial marriage. By 2011, according to Gallup, 86 percent of Americans "approve[d] of marriage between blacks and whites," and the matter has been settled. Instead, gay marriage is a big issue today; but in 50 years (and perhaps much less given current trends), it is likely to be as much a nonissue as interracial marriage is now. Another issue will exist, though, and it will divide those predisposed toward supporting new and those predisposed toward supporting traditional lifestyles. The evidence we have presented here says little about the coming and going of these issues (we leave that important topic to others), but says quite a bit about the type of person who ends up on one side or the other of whatever issue has been framed as a contrast between tradition and innovation.

Second, nothing in the empirical evidence or in our language should be taken to mean that one particular ideological stance is better or more natural than another. We know how the game is played, and some people will undoubtedly find an interpretation or a turn of phrase that reveals our deep hostility toward liberals or toward conservatives. This is as certain as it is depressing, leaving us to appeal somewhat forlornly to the strongly predisposed to beat down the instinct to be defensive, even if our terminology has been off-putting on occasion. The evidence of a biological and deep psychological substrate explains why so much variation in political temperaments exists today, yesterday, and tomorrow. It does not say anything about one particular temperament being better than another.

Finally, we make one last plea to think probabilistically. We have come a long way since illustrating how correlations are measured. We now have seen that correlations between physiological and psychological traits and political orientations are important; significant within studies; and consistent across studies, countries, and samples—but also modest in strength. This means that there are surprising differences between liberals and conservatives on an incredible range of traits, many not obviously related to politics, but it also means that exceptions are plentiful. Not all people who tend to prefer solutions that are characterized as conservative live in Conservaton and not all people with liberal views make a home in Liberalville. Only *on average* are conservatives more likely to behave like Conservatonians and liberals like Liberalites. To say otherwise would be to engage in stereotyping. Mustering one, two, or even a hundred cases that run contrary to the reported pattern does nothing to contradict the general relationships we have examined. Think probabilistically and do not pretend the research is claiming more than it is.

Conclusion: You Are Special . . . but Don't Let It Go to Your Head

You were born with a unique genetic package. This package was immediately modified by prenatal and early postnatal forces, and further modified by a wide range of environmental influences during development and beyond. These sources of influence combined into dispositional tendencies that affect your behavioral and attitudinal responses to whatever situations the world presents to you. These tendencies are inertial; they structure your attitudes and behaviors but do not predetermine them. Politics might seem as though it should be immune from such predispositional forces, but in this book we have dissected a rapidly growing corpus of research indicating that this is simply not the case. The political diversity that springs from differences in predispositions will never go away, and in a surprising number of cases it can meaningfully be arrayed on a spectrum that runs from supporters of tradition to supporters of innovation (conservatives and liberals, respectively, to use phrases that are common in the modern United States), with many possible positions in between the two. The conflict resulting from political diversity is often debilitating and occasionally even bloody. Still, if you accept that your political views are imbued less with majestic rationality than primal biology, that they bubble up from within rather than get passed down from on high, and if you recognize that predispositions affect how people perceive, process, and experience the world, you will have learned something valuable not just about yourself but about.

Notes

1   Shapiro and Bloch-Elkon, "Do Facts Speak for Themselves? Partisan Disagreement as a Challenge to Democratic Theory."

2   Kriegel, *The French Communists: Profile of a People*. See also Fejto, *The French Communist Party and the Crisis of International Communism*.

3   Delli Carpini and Keeter, *What Americans Know about Politics and Why It Matters*.

4   Klein, "American Misperceptions of Foreign Aid."

5   Delli Carpini and Keeter, *What Americans Know about Politics and Why It Matters*, 98. For an interesting distinction between people who are "confidently" wrong and those who are just wrong, see Kuklinski et al., "Misinformation and the Currency of Democratic Citizenship."

6   Redlawsk et al., "The Affective Tipping Point: Do Motivated Reasoners Ever 'Get It'?"

7   Frenda et al., "False Memories of Fabricated Political Events."

8   Quoted in Witchel, "Life after 'Sex.'" Available at http://www.nytimes.com/2012/01/22/magazine/cynthia-nixon-wit.html?pagewanted = all&_r = 0.

9   Quoted in Kaplan, "Cynthia Nixon Says She's Gay by 'Choice.' Is It Really a Choice?" Available at http://articles.latimes.com/2012/jan/25/news/la-heb-cynthia-nixon-gay-by-choice-20120125.

10   Wilson, *Sociobiology: The New Synthesis*.

11   Segerstrale, *Defenders of the Truth*.

12   Allen et al., "Against Sociobiology." Available at http://www.nybooks.com/articles/archives/1975/nov/13/against-sociobiology/.

13   For a comprehensive account of the sociobiology controversy see Segerstrale, *Defenders of the Truth*.

14   Segerstrale, *Defenders of the Truth*, 23.

15   Murray and Herrnstein, *The Bell Curve*. Herrnstein, along with educational psychologist Arthur Jensen, published some controversial arguments about IQ and race in the 1970s, not only long before *The Bell Curve*, but several years before the sociobiology controversy engulfed Wilson.

16   Perhaps the best "liberal" case—certainly one of the most readable and accessible—against the concept of IQ in general and the biological determinism of intelligence is Gould, *The Mismeasure of Man*.

17   Pearson, *National Life from the Standpoint of Science*.

18   The concept is similar to what psychologists sometimes call motivated cognition, but we opt for predisposition so that we can give more emphasis to the biological bases for the biases.

19   Porges, "Neuroception: A Subconscious System for Detecting Threat and Safety."

20   Lodge and Taber, *The Rationalizing Voter*.

21   See, for example, Haidt, *The Righteous Mind*.

22   Religioustolerance.org, "Causes of Homosexuality and Other Sexual Orientations: Public Opinion Polls," available at http://www.religioustolerance.org/hom_caus2.htm; and Gallup, "Tolerance for Gay Rights at High-Water Mark," available at http://www.gallup.com/poll/27694/tolerance-gay-rights-highwater-mark.aspx.

23   Sheldon et al., "Beliefs about the Etiology of Homosexuality and about the Ramifications of Discovering Its Possible Genetic Origin."

24   Langstrom et al., "Genetic and Environmental Effects on Same-Sex Sexual Behavior: A Population Study of Twins in Sweden"; LeVay, "A Difference in Hypothalmic Structure between Heterosexual and Homosexual Men"; and Hines, "Gender Development and the Human Brain."

25   Haidt, *The Righteous Mind*, 127.

26   Lakoff, *Moral Politics: How Liberals and Conservatives Think.*

27   Vedantam, *The Hidden Brain*, 6.

28   Madison, "Federalist #10."

29   A good introduction to the basic theory and aims of deliberative democracy is Fishkin, *Democracy and Deliberation: New Directions for Democratic Reform.*

30   For example, see Denver et al., "Fishkin and the Deliberative Opinion Poll: Lessons from a Study of the Granada 500 Television Program"; Barabas, "How Deliberation Affects Policy Opinions"; and Sulkin and Simon, "Habermas in the Lab: A Study of Deliberation in an Experimental Setting."

31   Hibbing and Theiss-Morse, *Stealth Democracy.*

The Left/Right 20 Questions Game

The 5 Questions from *Hardwired*:[1]

1. Could you slap your father in the face (with his permission) as part of a comedy skit?
   a) Yes   b) No
2. When you go to work in the morning, do you often leave a mess in your apartment or house?
   a) Yes   b) No
3. Which lesson is more important to teach to children?
   a) Kindness   b) Respect
4. Do you get bored by abstract ideas and theoretical discussions?
   a) Yes   b) No
5. Think about this carefully for 15 seconds: "Cleanliness is next to Godliness." Which answer is closer to your current thoughts?
   a) Okay . . . makes sense.   b) What???

Which item from each pair comes closest to describing you?[2]

6. a) Eccentric   b) Conventional
7. a) Decisive   b) Flexible
8. a) Open-minded   b) Moralistic
9. a) Imaginative   b) Practical
10. a) Simple   b) Complex

If forced to pick only one from each pair, which would you prefer?

11. a) Small towns   b) Big cities
12. a) Romantic movies   b) Comedies
13. a) Country music   b) Classical music
14. a) Motorcycle   b) SUV
15. a) Book about sports   b) Book about music

Are you closer to agreeing or disagreeing with the following statements?[3]

16. Some people are just more worthy than others.
    a) Agree   b) Disagree
17. If people were treated more equally, we would have fewer problems in this country.
    a) Agree   b) Disagree
18. To get ahead in life, it is sometimes necessary to step on others.
    a) Agree   b) Disagree
19. In an ideal world, all nations would be equal.
    a) Agree   b) Disagree
20. It's probably a good thing that certain groups are at the top and other groups are at the bottom.
    a) Agree   b) Disagree

Score +1 for each "a" answer on questions 4, 5, 7, 10, 11, 13, 15, 16, 18, and 20 and +1 for each "b" answer on questions 1, 2, 3, 6, 8, 9, 12, 14, 17, 19. A total score of 0 is the farthest left, a score of 20 is the farthest right, and a score of 10 is middle of the road.

Notes

[1]   Christine Lavin, John Alford, John Hibbing, Jeff Mondak, and Gene Weingarten. (©2009). *Hardwired.*

[2]   Dana R. Carney, John T. Jost, Samuel D. Gosling, and Jeff Potter. (2008). "The Secret Lives of Liberals and Conservatives: Personality Profiles, Interaction Styles, and the Things They Leave Behind." *Political Psychology, 29(6),* pp. 807–840.

[3]   Felicia Pratto, James Sidanius, Lisa M. Stallworth, and Bertram F. Malle. (1994). "Social Dominance Orientation: A Personality Variable Predicting Social and Political Attitudes." *Journal of Personality and Social Psychology, 67(4),* pp. 741–763.

BIBLIOGRAPHY

Aaron, A., Badre, D., Brett, M., Cacioppo, J., Chambers, C., Cools, R., Engel, S., D'Esposito, M., Frith, C., Harmon-Jones, E., Jonides, J., Knutson, B., Phelps, L., Poldrack, R., Wager, T., Wagner, A., and Winkielman, P. (2007, 14 November). "Politics and the Brain." *The New York Times*.

Ackerman, J. M., Nocera, C. C., and Bargh, J. A. (2010). "Incidental Haptic Sensations Influence Social Judgments and Decisions." *Science, 328(5986),* 1712–1715.

Adorno, T. W., Frenkel-Brunswik, E., Levinson, D. J., and Sanford, R. N. (1950). *The Authoritarian Personality*. New York, NY: Harper and Row.

Alford, J. R., Funk, C. L., and Hibbing, J. R. (2005). "Are Political Orientations Genetically Transmitted?" *American Political Science Review, 99(2),* 153–167.

Alford, J. R., Hatemi, P., Hibbing, J. R., Martin, N., and Eaves, L. (2011). "The Politics of Mate Choice." *Journal of Politics, 73,* 362–379.

Allen, E., Beckwith, B., Beckwith, J., Chorover, S., and Culver, D. (1975, 7 August). "Against 'Sociobiology.'" *The New York Times Review of Books*.

Altemeyer, R. 1981. *Right-Wing Authoritarianism*. Winnipeg, Canada: University of Manitoba Press.

Amodio, D., Jost, J. T., Master, S., and Yee, C. (2007). "Neurocognitive Correlates of Liberalism and Conservatism." *Nature Neuroscience, 10,* 1246–1247.

Andersen, S., Ertac, S., Gneezy, U., Hoffman, M., and List, J. A. (2011). "Stakes Matter in Ultimatum Games." *American Economic Review, 101,* 3427–3439.

Antonakis, J., and Dalgas, O. (2009). "Predicting Elections: Child's Play!" *Science, 323(5918),* 1183.

Armstrong, J. (1955). "The Enigma of Senator Taft and American Foreign Policy." *Review of Politics, 17,* 206–231.

Arnhart, L. (2005). *Darwinian Conservatism*. Charlottesville, VA: Imprint Academic Publishing.

Bannerman, L. (2008, 7 October). "The Camp That 'Cures' Homosexuality." *The Times* (London).

Barabas, J. (2004). "How Deliberation Affects Policy Opinions." *American Political Science Review, 98,* 687–701.

Barker, E. (1958). *The Politics of Aristotle*. London, UK: Oxford University Press.

Baron-Cohen, S. (2003). *The Essential Difference: Men, Women, and the Extreme Male Brain*. New York, NY: Basic Books.

Barville, J. (2009, 1 October). "Pair Were Switched at Birth." *The Spokesman-Review* (Spokane, WA).

Bayliss, A. P., di Pellegrino, G., and Tipper, S. P. (2005). "Sex Differences in Eye Gaze and Symbolic Cueing of Attention." *Quarterly Journal of Experimental Psychology, 58,* 631–650.

Bayliss, A. P., and Tipper, S. P. (2005). "Gaze and Arrow Cueing of Attention Reveals Individual Differences along the Autism Spectrum as a Function of Target Context." *British Journal of Psychology, 96,* 95–114.

BBC Radio Four. "Colin Firth: An Opportunity to Explore." Retrieved at http://news.bbc.co.uk/today/hi/today/newsid_9323000/9323470.stm

Bell, A. M. (2007). "Future Directions in Behavioural Syndromes Research." *Proceedings of the Royal Society, 247(1611),* 755–761.

Bell, D. (1960). *The End of Ideology: On the Exhaustion of Political Ideas in the Fifties.* Cambridge, MA: Harvard University Press.

Bell, E., Schermer, J. A., and Vernon, P. A. (2009). "The Origins of Political Attitudes and Behaviours: An Analysis Using Twins." *Canadian Journal of Political Science, 42(4),* 855–879.

Benjamin, D., Cesarini, D., Matthijs, J., Daws, C., Kellinger, P., Manusson, P., Chabris, C., Conley, D., Laibson, D., Johannesson, M., and Visscher, P. (2012). "Genetic Architecture of Economic and Political Preferences." *Proceedings of the National Academy of Sciences, 109(21),* 8026–2031.

Bennett, C., Baird, A., Miller, M., and Wolford, G. (2012). "Neural Correlates of Interspecies Perspective Taking in the Post-Mortem Atlantic Salmon: An Argument for Proper Multiple Comparisons Correction." *Journal of Serendipitous and Unexpected Results, 1,* 1–5.

Berger, J., Meredith, M., and Wheeler, S. C. (2008). "Contextual Priming: Where People Vote Affects How They Vote." *Proceedings of the National Academy of Sciences of the United States of America, 105(26),* 8846–8849.

Block, J., and Block, J. H. (2006). "Nursery School Personality and Political Orientation Two Decades Later." *Journal of Research in Personality, 40,* 734–749.

Bobbio, N. (1996). *Left and Right.* Chicago: University of Chicago Press.

Bohnet, I., and Frey, B. (1999). "Social Distance and Other-Regarding Behavior in Dictator Games: Comment." *American Economic Review, 89(1),* 335–339.

Bouchard, T. J., and McGue, M. (2003). "Genetic and Environmental Influences on Human Psychological Differences." *Developmental Neurobiology, 54(1),* 4–45.

Bouchard, T. J., Segal, N. L., Tellegen, A., McGue, M., Keyes, M., and Krueger, R. F. (2004). "Genetic Influence on Social Attitudes: Another Challenge to Psychology from Behavior Genetics." In L. DiLalla (Ed.), *Behavior Genetic Principles: Development, Personality, and Psychopathology.* Washington, D.C.: American Psychological Association Press.

Bradley, M. M., and Lang, P. J. (2007). "The International Affective Picture System (IAPS) in the Study of Emotion and Attention." In Coan, J. A., and Allen, J. J. B. (Eds.), *Handbook of Emotion Elicitation and Assessment.* Oxford, UK: Oxford University Press.

Bull, R., and Hawkes, C. (1982). "Judging Politicians by Their Faces." *Political Studies, 30,* 95–101.

Bull, R., Jenkins, M., and Stevens, J. (1983). "Evaluations of Politicians' Faces." *Political Psychology, 4(4),* 713–716.

Buller, D. J. (2006). *Adapting Minds.* Cambridge, MA: MIT Press.

Burns, J. M., and Swerdlow, R. H. (2003). "Right Orbitofrontal Tumor with Pedophilia Symptom and Constructional Apraxia Sign." *Archives of Neurology, 60,* 437–440.

Buss, David M., and Greiling, Heidi. (1999). "Adaptive Individual Differences." *Journal of Personality 67(2),* 209–243.

Buswell, G. T. (1937). *How Adults Read.* Chicago, IL: University of Chicago Press.

Cacioppo, J. T., Petty, R. E., and Tassinary, L. G. (1989). "Social Psychophysiology: A New Look." *Advances in Experimental Social Psychology, 22,* 39–55.

Camerer, C. F., Loewenstein, G., and Prelec, D. (2005). "Neuroeconomics: How Neuroscience Can Inform Economics." *Journal of Economic Literature, 43(1),* 9–64.

Cameron, L. A. (1999). "Raising the Stakes in the Ultimatum Game: Experimental Evidence from Indonesia." *Economic Inquiry, 37,* 47–59.

Caprara, G. V., Barbaranelli, C., and Zimbardo, P. G. (1999). "Personality Profiles and Political Parties." *Political Psychology, 20(1),* 175–197.

Carney, D., Jost, J., Gosling, S., and Potter, J. (2008). "The Secret Lives of Liberals and Conservatives: Personality Profiles, Interaction Styles, and the Things They Leave Behind." *Political Psychology, 29(6),* 807–840.

Carraro, L., Castelli, L., and Macchiella, C. (2011). "The Automatic Conservative: Ideology-Based Attentional Asymmetries in the Processing of Valenced Information." *PLoS ONE, 6(11).*

Cashdan, E. (1980). "Egalitarianism among Hunters and Gatherers." *American Anthropologist, 82,* 116–120.

Castelli, L., and Carraro, L. (2011). "Ideology Is Related to Basic Cognitive Processes Involved in Attitude Formation." *Journal of Experimental Social Psychology, 47(5),* 1013–1016.

Castiello, U., Zucco, A., Parma, V., Ansuini, C., and Tirindelli, R. (2006). "Cross-Modal Interactions between Olfaction and Vision When Grasping." *Chemical Senses, 31,* 665–671.

Cesarini, D., Johannesson, M., and Oskarsson, S. (n.d.). "Pre-Birth Factors and Voting: Evidence from Swedish Adoption Data." Unpublished manuscript.

Charness, G., and Gneezy, U. (2008). "What's in a Name? Anonymity and Social Distance in Dictator and Ultimatum Games." *Journal of Economic Behavior and Organization, 68(1),* 29–35.

Charney, E. (2008). "Genes and Ideologies." *Perspectives on Politics, 6(2),* 299–319.

Charney, E., and English, W. (2012). "Candidate Genes and Political Behavior." *American Political Science Review, 106(1),* 1–34.

Chirumbolo, A., Areni, A., and Sensales, G. (2004). "Need for Cognitive Closure and Politics: Voting, Political Attitudes and Attributional Style." *International Journal of Psychology, 39(4),* 245–253.

Cochran, G., and Harpending, H. (2009). *The 10,000 Year Explosion.* New York, NY: Basic.

Coleman, J. (2000). *A History of Political Thought.* Oxford, UK: Blackwell Publishers.

Cook, M., and Mineka, S. (1989). "Observational Conditioning of Fear to Fear-Relevant Versus Fear-Irrelevant Stimuli in Rhesus Monkeys." *Journal of Abnormal Psychology, 98(4),* 448–459.

Conover, P. J., and Feldman, S. (1981). "The Origins and Meaning of Liberal/Conservative Self-Identifications." *American Journal of Political Science, 25(4),* 617–645.

Converse, P. E. (1964). "The Nature of Belief Systems in Mass Publics." In D. E. Apter (Ed.), *Ideology and Discontent.* New York, NY: Free Press.

Cornford, F. (1957). *From Religion to Philosophy.* New York, NY: Harper.

Cosmides, L., and Tooby, J. (2009). "What Is Evolutionary Psychology?" Retrieved at http://www.cep.ucsb.edu/155/WEP15508.pdf

Costa, P. T., and McCrae, R. R. (1992). *NEO PI-R. Professional Manual.* Odessa, FL: Psychological Assessment Resources, Inc.

Daly, M., and Wilson, M. (1988). *Homicide.* New Brunswick, NJ: Transaction Publishers.

Damasio, A. R., Everitt, B. J., and Bishop, D. (1996). "The Somatic Marker Hypothesis and the Possible Functions of the Prefrontal Cortex [and Discussion]." *Philosophical Transactions of the Royal Society, 351(1346),* 1413–1420.

Dambrun, M., Despres, G., and Guimond, S. (2003). "On the Multifaceted Nature of Prejudice: Psychophysiological Responses to Ingroup and Outgroup Ethnic Stimuli." *Current Research in Social Psychology, 8,* 187–206.

Danziger, S., Levav, J., and Avnaim-Pesso, L. (2011). "Extraneous Factors in Judicial Decisions." *Proceedings of the National Academy of Sciences of the United States of America, 108(17),* 6889–6892.

Dawson, M., Schell, A., and Filion, D. (2007). "The Electrodermal System." In J. Cacioppo, L. Tassinary, and G. Bertson (Eds.), *Handbook of Psychophysiology*. New York: Cambridge University Press.

Delli Carpini, M. X., and Keeter, S. (1996). *What Americans Know about Politics and Why It Matters.* New Haven, CT: Yale University Press.

Demir, E., and Dickson, B. J. (2005). "Fruitless Splicing Specifies Male Courtship Behavior in Drosophila." *Cell, 121(5),* 785–794.

Denver, D., Hands, G., and Jones, B. (1995). "Fishkin and the Deliberative Opinion Poll: Lessons from a Study of the Granada 500 Television Program." *Political Communication, 12,* 147–156.

Deppe, K. D., Stoltenberg, S. F., Smith, K. B., and Hibbing, J. R. (2013). "Candidate Genes and Voter Turnout: Further Evidence on the Role of 5-HTTLPR." *American Political Science Review, 107(2),* 375–81.

De Quervain, D. J.-F., Fischbacher, U., Treyer, V., Schellhammer, M., Schnyder, U., Buck, A., and Fehr, E. (2004). "The Neural Basis of Altruistic Punishment." *Science, 305(5688),* 1254–1258.

De Waal, F. (1996). *Good Natured: The Origins of Right and Wrong in Humans and Other Animals.* Cambridge, MA: Harvard University Press.

Dewy, J., and Kallen, H. M. (Eds.). (1941). *The Bertrand Russell Case.* New York, NY: The Viking Press.

Diamond, S. (1936). "The Co-Ordination of Erich Jaensch." *Science & Society, 1(1),* 106–114.

Digman, J. (1990). "Personality Structure: Emergence of the Five-Factor Model." *American Review of Psychology, 41,* 417–440.

Ditto, P. H., and Lopez, D. F. (1992). "Motivated Skepticism: Use of Differential Decision Criteria for Preferred and Nonpreferred Conclusions." *Journal of Personality and Social Psychology, 63(4),* 568–584.

Dodd, M. D., Balzer, A., Jacobs, C. M., Gruszczynski, M., Smith, K. B., and Hibbing, J. R. (2012). "The Political Left Rolls with the Good and the Political Right Confronts the Bad: Connecting Physiology and Cognition to Preferences." *Philosophical Transactions of the Royal Society, 367(1589),* 640–649.

Dodd, M. D., Hibbing, J. R., and Smith, K. B. (2011). "The Politics of Attention: Gaze-Cueing Effects Are Moderated by Political Temperament." *Attention, Perception, and Psychophysics, 73(1),* 24–29.

Dollinger, S. (2007). "Creativity and Conservatism." *Personality and Individual Differences, 43(5),* 1025–1035.

Dreber, A., Apicella, C. L., Eisenberg, D. T. A., Garcia, J. R., Zamore, R. S., Lum, J. K., and Campbell, B. (2009). "The 7R Polymorphism in the Dopamine Receptor D4 Gene (DRD4) Is Associated with Financial Risk Taking in Men." *Evolution and Human Behavior, 30(2),* 85–92.

Driver, J., Blankenburg, F., Bestmann, S., and Ruff, C. C. (2010). "New Approaches to the Study of Human Brain Networks Underlying Spatial Attention and Related Processes." *Experimental Brain Research, 206(2),* 153–162.

Durkheim, E. (1893/1997). *Division of Labor in Society.* New York, NY: Free Press.

Eagleman, D. (2011). *Incognito: The Secret Lives of the Brain.* New York, NY: Vintage Books.

Edwards, K., and Smith, E. E. (1996). "A Disconfirmation Bias in the Evaluation of Arguments." *Journal of Personality and Social Psychology, 7(1),* 5–24.

Eisner, M. (2001). "Modernization, Self-Control, and Lethal Violence: The Long-Term Dynamics of European Homicide Rates." *British Journal of Criminology, 41,* 618–638.

Ekman, P., Davidson, R. J., and Friesen, W. V. (1990). "The Duchenne Smile: Emotional Expression and Brain Physiology: II." *Journal of Personality and Social Psychology, 58(2),* 342–353.

Emerson, R. W. (1987). *The Essays of Ralph Waldo Emerson.* Cambridge, MA: Harvard University Press. (Original work published in 1841)

Faulkner, W. (1936). *Absalom, Absalom!* New York: Random House.

Federico, C. M., Golec, A., and Dial, J. L. (2005). "The Relationship between the Need for Closure and Support for Military Action against Iraq: Moderating Effects of National Attachment." *Personality and Social Psychology Bulletin, 31(5),* 621–632.

Feist, G. J., and Brady, T. R. (2004). "Openness to Experience, Non-Conformity, and the Preference for Abstract Art." *Empirical Studies of the Arts 22(1),* 77–89.

Fejto, F. (1967). *The French Communist Party and the Crisis of International Communism.* Cambridge, MA: MIT Press.

Figueredo, A. J., Gladden, P., Vasquez, G., Wolf, P. S. A., and Jones, D. N. "Evolutionary Theories of Personality." (2009). In P. J. Corr and G. Matthews (Eds.), *Cambridge Handbook of Personality Psychology: Part IV: Biological Perspectives* (pp. 265–274). Cambridge UK: Cambridge University Press.

Fiorina, M. P., Abrams, S. J., and Pope, J. C. (2005). *Culture War? The Myth of a Polarized America.* New York, NY: Pearson Longman.

Fisher, R. A. (1958). *The Genetical Theory of Natural Selection* (2nd ed). New York, NY: Dover.

Fishkin, J. (1991). *Democracy and Deliberation: New Directions for Democratic Reform.* New Haven, CT: Yale University Press.

Fowler, J. H., Baker, L. A., and Dawes, C. T. (2008). "Genetic Variation in Political Participation." *American Political Science Review, 102(2),* 233–248.

Fowler, J. H., and Dawes, C. T. (2008). "Two Genes Predict Voter Turnout." *Journal of Politics, 70(3),* 579–594.

Franken, A. (1999). *Rush Limbaugh Is a Big Fat Idiot.* New York, NY: Dell Publishing.

Freeman, D. (1983). *Margaret Mead and Samoa: The Making and Unmaking of an Anthropological Myth.* Cambridge, MA: Harvard University Press.

Frenda, S. J., Knowles, E. D., Saletan, W., and Loftus, E. F. (2013). "False Memories of Fabricated Political Events." *Journal of Experimental Social Psychology 49,* 280–286.

Friesen, C. K., and Kingstone, A. (1998). "The Eyes Have It! Reflexive Orienting Is Triggered by Nonpredictive Gaze." *Psychonomic Bulletin and Review, 5(3),* 490–495.

Friesen, C. K., Ristic, J., and Kingstone, A. (2004). "Attentional Effects of Counterpredictive Gaze and Arrow Cues." *Journal of Experimental Psychology, Human Perception and Performance, 30(2),* 319–329.

Fromm, E. (1941). *Escape from Freedom.* New York, NY: Holt, Rinehart, and Winston.

Fujita, F., and Diener, E. (2005). "Life Satisfaction Set Point: Stability and Change." *Journal of Personality and Social Psychology, 88(1),* 158–164.

Fukuyama, F. (2006, 1992). *The End of History and the Last Man.* New York, NY: Free Press.

———. (1995). *Trust: The Social Virtues and the Creation of Prosperity.* New York, NY: Free Press.

Funk, C., Smith, K. B., Alford, J. R., Hibbing, M. V., Eaton, N., Krueger, R., Eaves, L., and Hibbing, J. R. (2013). "Genetic and Environmental Transmission of Political Orientations." *Political Psychology, 34,* doi: 10.1111/j.1467-9221.2012.00915.x

Furnham, A., and Avison, M. (1997). "Personality and Preferences for Surreal Art." *Personality and Individual Differences, 23(6),* 923–935.

Furnham, A., and Walker, J. (2001). "The Influence of Personality Traits, Previous Experience of Art, and Demographic Variables on Artistic Preference." *Personality and Individual Differences, 31(6),* 997–1017.

Gabler, N. (2012, March). "The Weird World of Biopolitics." *Playboy, 76,* 127–130.

Galdi, S., Arcuri, L., and Gawronski, B. (2008). "Automatic Mental Associations Predict Future Choices of Undecided Decision-Makers." *Science, 321,* 1100–1102.

Gallup. (2007). "Tolerance for Gay Rights at High-Water Mark." Retrieved at http://www.gallup.com/poll/27694/tolerance-gay-rights-highwater-mark.aspx

Garcia, J., Kimeldorf, D. J., and Koelling, R. A. (1955). "Conditioned Aversion to Saccharin Resulting from Exposure to Gamma Radiation." *Science, 122,* 157–158.

Gazzaniga, M. (2011). *Who's in Charge?: Free Will and the Science of the Brain.* New York, NY: Harper Collins.

Gerber, A. S., Green, D. P., and Schachar, R. (2003). "Voting May Be Habit-Forming: Evidence from a Randomized Field Experiment." *American Journal of Political Science, 47(3),* 540–550.

Gerber, A. S., Huber, G. A., Doherty, D., Dowling, C. M., and Ha, S. E. (2010). "Personality and Political Attitudes: Relationships across Issue Domains and Political Contexts." *American Political Science Review, 104(1),* 111–133.

Gillies, J., and Campbell, S. (1985). "Conservatism and Poetry Preferences." *British Journal of Social Psychology 24(3),* 223–227.

Gladwell, M. (2008). *Outliers: The Story of Success.* New York, NY: Little, Brown, and Company.

Goldberg, L. R. (1993). "The Structure of Phenotypic Personality Traits." *American Psychologist, 48(1),* 26–34.

Goldman, J. G. (2010). "Man's New Best Friend? A Forgotten Russian Experiment in Fox Domestication." *Scientific American* blog. Retrieved at http://blogs.scientificamerican.com/guest-blog/2010/09/06/mans-new-best-friend-a-forgotten-russian-experiment-in-fox-domestication/

Goldstein, D. B. (2009). "Common Genetic Variation and Human Traits." *New England Journal of Medicine, 360(17),* 1696–1698.

Golec, A. (2002). "Need for Cognitive Closure and Political Conservatism: Studies on the Nature of the Relationship." *Polish Psychological Bulletin, 33(4),* 5–12.

Golec, A., Cislak, A., and Wesolowska, E. (2010). "Political Conservatism, Need for Cognitive Closure, and Intergroup Hostility." *Political Psychology, 31(4),* 521–541.

Gould, S. J. (1981). *The Mismeasure of Man.* New York, NY: W.W. Norton.

Grahm, J., Haidt, J., and Nosek, B. A. (2009). "Liberals and Conservatives Rely on Different Sets of Moral Foundations." *Journal of Personality and Social Psychology, 95(5),* 1029–1046.

Gross, J. J., and John, O. P. (1995). "Facets of Emotional Expressivity: Three Self-Report Factors and their Correlates." *Personality and Individual Differences, 19(4),* 555–568.

Gross, N. (2012, 4 March). "The Indoctrination Myth." *The New York Times.*

Haidt, J. (2012). *The Righteous Mind.* New York, NY: Pantheon Books.

Haidt, J., and Graham, J. (2007). "When Morality Opposes Justice: Conservatives Have Moral Intuitions That Liberals May Not Recognize." *Social Justice Research, 20,* 98–116.

Haidt, J., Graham, J., and Joseph, C. (2009). "Above and below Left-Right: Ideological Narratives and Moral Foundations." *Psychological Inquiry, 20,* 110–119.

Haidt, J., and Hersh, M. (2001). "Sexual Morality: The Cultures of Conservatives and Liberals." *Journal of Applied Social Psychology, 310,* 181–221.

Hammock, E. A. D., and Young, L. J. (2005). "Microsatellite Instability Generates Diversity in Brain and Sociological Traits." *Science, 308(5728),* 1630–1634.

Harlow, H. F., Dodsworth, R. O., and Harlow, M. K. (1965). "Total Social Isolation in Monkeys." *Proceedings of the National Academy of Sciences of the United States of America, 54(1),* 90–97.

Harpending, H., and Cochran, G. (2002). "In Our Genes." *Proceedings of the National Academy of Sciences of the United States of American, 99(1),* 10–12.

Harrison, N. A., Gray, M. A., Giananors, P. J., and Critchley, H. D. (2010). "The Embodiment of Emotional Feelings in the Brain." *Journal of Neuroscience, 30,* 12878–12884.

Haston, R. (2009). *So You Married a Conservative: A Stone Age Explanation of Our Differences, a New Path Towards Progress.* Retrieved at http://www.politicalspecies.com/SYMAC%20 Excerpt.pdf

Hatemi, P. K., Gillespie, N. A., Eaves, L. J., Maher, B. S., Webb, B. T., Heath, A. C., Medland, S. E., Smyth, D. C., Beeby, H. N., Gordon, S. D., Montgomery, G. W., Zhu, G., Byrne, E. M., and Martin, N. G. (2011). "A Genome-Wide Analysis of Liberal and Conservative Political Attitudes." *Journal of Politics, 73(1),* 271–285.

Hatemi, P. K., Medland, S. E., Morley, K. I., Heath, A. C., and Martin, N. G. (2007). "The Genetics of Voting: An Australian Twin Study." *Behavior Genetics, 37,* 435–448.

Helzer, E. G., and Pizarro, D. A. (2011). "Dirty Liberals! Reminders of Physical Cleanliness Influence Moral and Political Attitudes." *Psychological Science, 22(4),* 517–522.

Henrich, J., Boyd, R., Bowles, S., Camerer, C., Fehr, E., Gintis, H., and McElreath, R. (2001). "In Search of Homo Economicus: Behavioral Experiments in 15 Small-Scale Societies." *American Economic Review, 91(2),* 73–78.

Hetherington, M., and Weiler, J. (2009). *Authoritarianism and Polarization in American Politics.* New York, NY: Cambridge University Press.

Heywood, A. (2007). *Political Ideologies.* New York, NY: Palgrave Macmillan.

Hibbing, J. R., Smith, K. B., and Alford, J. R. (2014). Differences in Negativity Bias Underlie Variations in Political Ideology." *Behavioral and Brain Sciences, 37,* in press.

Hibbing, J. R., and Theiss-Morse, E. (2002). *Stealth Democracy.* New York, NY: Cambridge University Press.

Hines, M. (2011). "Gender Development and the Human Brain." *Annual Review of Neuroscience, 24,* 69–88.

Hirsch, J. B., DeYoung, C. G., Xu, X., and Peterson, J. B. (2010). "Compassionate Liberals and Polite Conservatives: Associations of Agreeableness with Political Ideology and Moral Values." *Personality and Social Psychology Bulletin, 36(5), 655–664.*

Hobbes, T. (1651/2010). *Leviathan: Or the Matter, Forme, and Power of a Common-wealth Ecclesiastical and Civill.* I. Shapiro (Ed.). New Haven, CT: Yale University Press.

Hoffman, E., and Spitzer, M. (1985). "Entitlements, Rights, and Fairness: An Experimental Examination of Subjects' Concepts of Distributive Justice." *Journal of Legal Studies, 14,* 259–299.

Horner, V., and Whiten, A. (2005). "Causal Knowledge and Imitation/Emulation Switching in Chimpanzees and Children." *Animal Cognition, 8,* 164–181.

Huber, G., and Malhotra, N. (2012). "Political Sorting in Social Relationships: Evidence from an Online Dating Community." Unpublished manuscript. Retrieved at http://xa.yimg.com/kq/ groups/17296918/440105617/name/Political+Sorting+in+Social+Relationships.pdf

Huddy, L., and Feldman, S. (2011). "Americans Respond Politically to 9/11: Understanding the Impact of the Terrorist Attacks and Their Aftermath." *American Psychologist, 66(6),* 455–467.

Huey, E. (1908). *The Psychology and Pedagogy of Reading.* Norwood, MA: Macmillan.

Hunch.com. "How Food Preferences Vary by Political Ideology." Retrieved at http://hunch.com/ media/reports/food/

————. "You Vote What You Eat: How Liberals and Conservatives Eat Differently." Retrieved at http://blog.hunch.com/?p=48884

Iacoboni, M., Freedman, J., Kaplan, J., Jamieson, K. H., Freedman, T., Knapp, B., and Fitzgerald, K. (2007, 11 November). "This Is Your Brain on Politics." *The New York Times.*

Ibanga, I. (2009, 14 May). "Switched at Birth: Women Learn the Truth 56 Years Later." *Good Morning America.*

Inbar, Y., Pizarro, D. A., and Bloom, P. (2009). "Conservatives Are More Easily Disgusted Than Liberals." *Cognition and Emotion, 23(4),* 714.

————. (2012). "Disgusting Smells Cause Decreased Liking of Gay Men." *Emotion, 12(1),* 23–27.

Jacobs, C., Hibbing, J. R., and Smith, K. B. (2012). "Carrying Your Heart (and Your Politics) on Your Face: Ideology and Facial Muscle Responses." Paper presented at the Midwest Political Science Association Meetings, Chicago, April 2012.

Jahoda, G. (1954). "Political Attitudes and Judgments of Other People." *Journal of Abnormal and Social Psychology, 49,* 331–334.

Jang, K., Livesley, W. J., and Vernon, P. A. (1966). "Heritability of the Big Five Personality Dimensions and Their Facets: A Twin Study." *Journal of Personality, 64,* 577–591.

Jencks, C. (1980). "Heredity, Environment, and Public Policy Reconsidered." *American Sociological Review, 45(5),* 723–736.

Jennings, M. K., and Neimi, R. G. (1968). "The Transmission of Political Values from Parent to Child." *American Political Science Review, 62(1),* 169–184.

Jimenez, G. C. (2009). *Red Genes, Blue Genes: Exposing Political Irrationality.* New York, NY: Autonomedia.

Johnson, B. R. (2010). "Division of Labor in Honeybees: Form, Function, and Proximate Mechanisms." *Behavioral Ecology and Sociobiology, 64(3),* 305–316.

Johnson, C. (2012). *The Information Diet.* Sebastopol, CA: O'Reilly Media, Inc.

Jost, J. T. (2006). "The End of the End of Ideology." *The American Psychologist, 61,* 651–670.

Jost, J. T., Glaser, J., Kruglanski, A. W., and Sulloway, F. J. (2003). "Political Conservatism as Motivated Social Cognition." *Psychological Bulletin, 129(3),* 339–375.

Jost, J. T., and Kruglanski, A. W. (1999). "Effects of Epistemic Motivation on Conservatism, Intolerance, and Other System Justifying Attitudes." In L. Thompson, D. M. Messick, and J. M Levine (Eds.), *Shared Cognition in Organizations: The Management of Knowledge.* Hillsdale, NJ: Lawrence Erlbaum Associates, Inc.

Kanai, R., Feilden, T., Firth, C., and Rees, G. (2011). "Political Orientations Are Correlated with Brain Structure in Young Adults." *Current Biology, 21,* 1–4.

Kaplan, K. (2012, 25 January). "Cynthia Nixon Says She's Gay by 'Choice.' Is It Really a Choice?" *The Los Angeles Times.*

Keeley, L. H. (1996). *War before Civilization.* Oxford, UK: Oxford University Press.

Kettlewell, H. B. D. (1959). "Darwin's Missing Evidence." *Scientific American, 200,* 48–53.

Klein, E. (2010). "American Misperceptions of Foreign Aid." *The Washington Post* Online. Retrieved at http://voices.washingtonpost.com/ezra-klein/2010/12/american_misperceptions_of_for.html

Klemmensen, R., Hatemi, P. K., Hobolt, S. B., Petersen, I., Skytthe, A., and Norgaard, A. S. (2012). "The Genetics of Political Participation, Civic Duty, and Political Efficacy across Cultures: Denmark and the United States." *Journal of Theoretical Politics, 24(3),* 409–427.

Kohn, G. C. (2008). Encyclopedia of Plague and Pestilence (3rd ed). New York, NY: Infobase Publishing.

Kosofsky, E. S., and Frank, A. (1995). "Shame in the Cybernetic Fold: Reading Silvan Tomkins." *Critical Inquiry, 21(2),* 496–522.

Kossowska, M., and Van Hiel, A. (2003). "The Relationship between Need for Closure and Conservative Beliefs in Western and Eastern Europe." *Political Psychology, 24(3)*, 501–518.

Krebs, J. (2009). "The Gourmet Ape: Evolution and Human Food Preferences." *The American Journal of Clinical Nutrition, 90*, 707S–711S.

Kriegel, A. (1972). *The French Communists: Profile of a People*. Chicago, IL: University of Chicago Press.

Kristol, I. (1977, 23 January). "Memoirs of a Trotskyist." *The New York Times Magazine*.

Kruglanski, A. W., Webster, S. M., and Klem, A. (1993). "Motivated Resistance and Openness to Persuasion in the Presence or Absence of Prior Information." *Journal of Personality and Social Psychology, 65(5)*, 861–876.

Kuhnen, C. M., and Chiao, J. Y. (2009). "Genetic Determinants of Financial Risk Taking." *PLoS ONE, 4(2)*, e4362.

Kuklinski, J. H., Quirk, P., Jerit, J. T., Schweider, D., and Rich, R. (2000). "Misinformation and the Currency of Democratic Citizenship." *Journal of Politics, 62*, 790–815.

Laertius, D. (Third century AD/1923) "The Lives and Opinions of Eminent Philosophers: Life of Aristotle." (C. D. Yonge, Trans.). Retrieved at http://classicpersuasion.org/pw/diogenes/dlaristotle.htm

Lakoff, G. (2002). *Moral Politics: How Liberals and Conservatives Think*. Chicago, IL: University of Chicago Press.

Lane, R. (1962). *Political Ideology: Why the American Common Man Believes What He Does*. New York: Free Press.

Lang, P. (1995). "The Emotion Probe: Studies of Motivation and Attention." *American Psychologist, 50*, 372–385.

Langstrom, N., Rahman, Q., Carlstrom, E., and Lichtenstein, P. (2010). "Genetic and Environmental Effects on Same-Sex Sexual Behavior: A Population Study of Twins in Sweden." *Archives of Sexual Behavior, 39*, 75–80.

Laponce, J. A. (1981). *Left and Right: The Topography of Political Perceptions*. Toronto, Canada: University of Toronto Press.

Lasch, C. (1991). *The True and Only Heaven*. New York, NY: Norton.

Leder, M. (2006). "What Makes a Stock Republican?" Slate. Retrieved at http://www.slate.com/articles/business/moneybox/2006/11/what_makes_a_stock_republican.html

LeVay, S. (1994). "A Difference in Hypothalmic Structure between Heterosexual and Homosexual Men." *Science, 253*, 1034–1037.

Liang, Z. S., Nguyen, T., Mattila, H. R., Rodriguez-Zas, S. L., Seeley, T. D., and Robinson, G. E. (2012). "Molecular Determinants of Scouting Behavior in Honey Bees." *Science, 335(6073)*, 1225–1228.

Lienesch, M. (1982). "Right-Wing Religion: Christian Conservatism as a Political Movement." *Political Science Quarterly, 97(3)*, 403–425.

Locke, J. (1690/2003). *Two Treatises of Government*. New Haven: Yale University Press.

Lodge, M., and Hamill, R. (1986). "A Partisan Schema for Political Information Processing." *American Political Science Review, 80(2)*, 505–520.

Lodge, M., and Taber, C. S. (2005). "The Automaticity of Affect for Political Leaders, Groups, and Issues: An Experimental Test of the Hot Cognition Hypothesis." *Political Psychology, 26(3)*, 455–482.

———. (2013). *The Rationalizing Voter*. Cambridge, UK: Cambridge University Press.

Lykken, D. (1998). "The Genetics of Genius." In A. Steptoe (Ed.), *Genius and the Mind*. Oxford, UK: Oxford University Press.

Macmillan, M. (2008). "Phineas Gage—Unravelling the Myth." *The Psychologist, 21(9)*, 828–831.

Madison, J. (1788/1961). "Federalist #10." In C. Rossiter (Ed.), *The Federalist Papers*. New York: New American Library.

Maestripieri, D., Roney, J. R., DeBias, N., Durante, K. M., and Spaepen, G. (2004). "Father Absence, Menarche, and Interest in Infants among Adolescent Girls." *Developmental Science, 7(5),* 560–566.

Maguire, E. A., Woollet, K., and Spiers, H. (2000). "London Taxi Drivers and Bus Drivers: A Structural MRI and Neuropsychological Analysis." *Hippocampus, 16,* 1091–1101.

Marcus, G. E., Sullivan, J. L., Theiss-Morse, E., and Wood, S. (1995). *With Malice Toward Some: How People Make Civil Liberties Judgments*. Cambridge, UK: Cambridge University Press.

Martin, J. (2001). "The Authoritarian Personality, 50 Years Later: What Lessons Are There for Political Psychology?" *Political Psychology, 22,* 1–26.

Martin, N. G., Eaves, L. J., Heath, A. C., Jardine, R., Feingold, L. M., and Eysenck, H. J. (1986). "Transmission of Social Attitudes." *Proceedings of the National Academy of Sciences of the United States of America, 83(12),* 4364–4368.

McCabe, K., Houser, D., Ryan, L., Smith, V., and Trouard, T. (2001). "A Functional Imaging Study of Cooperation in Two-Person Reciprocal Exchange." *Proceedings of the National Academy of Sciences of the United States of America, 98(20),* 11832–11835.

McClosky, H. (1958). "Conservatism and Personality." *The American Political Science Review, 52(1),* 27–45.

McDermott, R. (2011). "Hormones and Politics." In R. McDermott and P. Hatemi, P. (Eds.), *Man Is by Nature a Political Animal*. Chicago, IL: University of Chicago Press.

McLean, S. P., Garza, J. P., Wiebe, S. A., Dodd, M. D., Smith, K. B., Hibbing, J. R., and Espy, K. A. (2013). "Applying the Flanker Task to Political Psychology: A Research Note." *Political Psychology*.

Miklosi, A., Polgardi, R., Topal, J., and Csanyi, V. (1998). "Use of Experimenter-Given Cues in Dogs." *Animal Cognition, 1,* 113–121.

Milgram, S. (1974). *Obedience to Authority*. New York, NY: Harper and Row.

Mill, J. S. (1985). *On Liberty*. London, UK: Penguin Books. (Original work published in 1859)

Miller, J. (1977). *The Wolf by the Ears*. New York, NY: Free Press.

Mitchell, D. G., Hibbing, M. V., Smith, K. B., and Hibbing, J. R. (2013) "Side by Side, Worlds Apart: Liberals' and Conservatives' Distinct Perceptions of Political Reality." American Politics Research, 42, in press.

Muller-Doohm, S. (2005). *Adorno: A Biography*. (R. Livingstone, Trans.) Cambridge, UK: Polity Press.

Mondak, J. J. (2010). *Personality and the Foundations of Political Behavior*. New York, NY: Cambridge University Press.

Mondak, J. J., Hibbing, M. V., Canache, D., Seligson, M. A., and Anderson, M. R. (2010). "Personality and Civic Engagement: An Integrative Framework for the Study of Trait Effects on Political Behavior." *American Political Science Review, 104(1),* 85–110.

Mooney, C. (2005). *The Republican War on Science*. New York, NY: Basic Books.

———. (2012). *The Republican Brain: The Science of Why They Deny Science—and Reality*. Hoboken, NJ: Wiley.

Moore, C., and Dunham, P. J. (1995). *Joint Attention: Its Origins and Role in Development*. Hillsdale, NJ: Lawrence Erlbaum Associates, Inc.

Moran, R. (1981). *Knowing Right from Wrong: The Insanity Defense of Daniel McNaughtan*. New York, NY: Free Press.

Morris, D. H., Jones, M. E., Minouk, M. J. K., Ashworth, A, and Swerdlow, A. J. (2012). "Family Concordance for Age at Menarche." *Pediatric Perinatal Epidemiology, 25(3),* 306–311.

Mueller, D. (1993). *The Public Choice Approach to Politics*. Aldershot, UK: Edward Elgar Publishing, Ltd.

Muir, W. M. (1995). "Group Selection for Adaptation to Multiple Hen Cages." *Poultry Science, 75,* 447–458.

Mulgan, R. G. (1974). "Aristotle's Doctrine That Man Is a Political Animal." *Hermes, 102,* 438–445.

Munro, G. D. (2010). "The Scientific Impotence Excuse: Discounting Belief-Threatening Scientific Abstracts." *Journal of Applied Social Psychology, 40(3),* 579-600.

Munro, G. D., Ditto, P., Lockhart, L., Fagerlin, A., Gready, M., and Peterson, E. (2002). "Biased Assimilation of Sociopolitical Arguments: Evaluating the 1996 U.S. Presidential Debate." *Basic and Applied Social Psychology, 24,* 15–26.

Murray, C., and Herrnstein, R. J. (1994). *The Bell Curve*. New York, NY: Free Press.

Mutz, D. (1998). *Impersonal Influence: How Perceptions of Mass Collectives Affect Political Attitudes*. Cambridge, UK: Cambridge University Press.

Napier, J. L. and Jost, J. T. (2008). "Why Are Conservatives Happier Than Liberals?" *Psychological Science, 19(6),* 565–572.

National Public Radio. (2005). "Frequently Asked Questions about Lobotomies." Retrieved at http://www.npr.org/templates/story/story.php?storyId=5014565

Neiman, J. (2012). "Phenylthiocarbamide Detection and Political Ideology." Paper presented at the annual meeting of the ISPP 35th Annual Scientific Meeting, Chicago, IL, July 2012.

———. (2012). "Political Ideology, Personality, and the Correlations with Tastes and Preferences for Music, Art, Literature, and Food." Presented at the annual meeting of the Great Plains Political Science Association, Hastings, NE, September 2012.

Niemi, R. G., and Jennings, M. K. (1991). "Issues and Inheritance in the Formation of Party Identification." *American Journal of Political Science, 35,* 970–988.

Noback, C., and Demarest, R. (1975). *The Human Nervous System: Basic Principles of Neurobiology*. New York, NY: McGraw-Hill.

Nowak, M. A., Tarnita, C. E., and Wilson, E. O. (2010). "The Evolution of Eusociality." *Nature, 446,* 1057–1062.

Olver, J. (2003). "Personality Traits and Personal Values: A Conceptual and Empirical Integration." *Personality and Individual Differences, 35,* 109–125.

O'Reilly, B. (2010). *Pinheads and Patriots: Where You Stand in the Age of Obama*. New York, NY: William Morrow.

Oxley, D., Smith, K. B., Alford, J., Hibbing, M., Miller, J., Scalora, M., Hatemi, P., and Hibbing, J. R. (2008). "Political Attitudes Vary with Physiological Traits." *Science, 321,* 1667–1670.

Pearson, K. (1919). *National Life from the Standpoint of Science* (2nd ed.). Cambridge, UK: Cambridge University Press.

Peterson, M. B., Sznycer, D., Sell, A., Cosimedes, L., and Tooby, J. (2013). "The Ancestral Logic of Politics: Upper Body Strength Regulates Men's Assertion of Self-Interest over Economic Redistribution." *Psychological Science,* doi: 10.1177/0956797612466415.

Pichot, P. (1984). "Centenary of the Birth of Hermann Rorschach." *Journal of Personality Assessment, 48,* 591–596.

Pinker, S. (2002). *The Blank Slate: The Modern Denial of Human Nature*. New York, NY: Viking.

———. (2011). *The Better Angels of Our Nature: Why Violence Has Declined*. New York, NY: Penguin Books.

Piurko, Y., Schwartz, S., and Davidov, E. (2011). "Basic Personal Values and the Meaning of Left-Right Political Orientations in 20 Countries." *Political Psychology, 32,* 537–561.

Plutzer, E. (2002). "Becoming a Habitual Voter: Inertia, Resources, and Growth in Young Adulthood." *American Political Science Review, 96(1)*, 41–56.

Porges, S. (2004). "Neuroception: A Subconscious System for Detecting Threat and Safety." *Zero to Three Journal, 24*, 9–24.

Pratto, F., Sidanius, J., Stallworth, L., and Malle, B. (1994). "Social Dominance Orientation: A Personality Variable Predicting Social and Political Attitudes." *Journal of Personality and Social Psychology, 67*, 741–763.

Prior, M. (2010). "You've Either Got It or You Don't? The Stability of Political Interest over the Life Cycle." *Journal of Politics, 72(3)*, 747–766.

Putnam, R. D. (1994). *Making Democracy Work: Civic Traditions in Modern Italy*. Princeton, NJ: Princeton University Press.

Pyysiäinen, I., and Hauser, M. (2010). "The Origins of Religion: Evolved Adaptation or By-Product?" *Trends in Cognitive Sciences, 14(3)*, 104–109.

Quinlan, Robert J. (2003). "Father Absence, Parental Care, and Female Reproductive Development." *Evolution and Human Behavior, 24(6)*, 376–390.

Raison, C. L., and Miller, A. H. (2003). "The Evolutionary Significance of Depression in Pathogen Host Defense (PATHOS-D)." *Molecular Psychiatry, 18*, 15–37.

Rankin, R. E., and Campbell, D. T. (1955). "Galvanic Skin Response to Negro and White Experimenters." *Journal of Abnormal and Social Psychology, 51*, 30–33.

Rawlings, D., Barrantes, I., Vidal, N., and Furnham, A. (2000). "Personality and Aesthetic Preference in Spain and England: Two Studies Relating Sensation Seeking and Openness to Experience to Liking for Paintings and Music." *European Journal of Personality, 14(6)*, 553–576.

Ray, J. J. (1974). "How Good Is the Wilson and Patterson Conservatism Scale?" *New Zealand Psychologist, 3*, 21–26.

———. (1989). "The Scientific Study of Ideology Is Too Often More Ideological Than Scientific." *Personality & Individual Differences, 10(3)*, 331–336.

Raymond, M., Pontier, D., Dufour, A. B., and Moller, A. P. (1996). "Frequency-Dependent Maintenance of Left-Handedness in Humans." *Proceedings of the Royal Society of London, B 263*, 1627–1633.

Redlawsk, D. P., Civettini, A. J. W., and Emmerson, K. M. (2010). "The Affective Tipping Point: Do Motivated Reasoners Ever 'Get It'?" *Political Psychology, 31(4)*, 563–593.

Religioustolerance.org. "Causes of Homosexuality and Other Sexual Orientations: Public Opinion Polls. Retrieved at http://www.religioustolerance.org/hom_caus2.htm

Risen, J. L., and Critcher, C. R. (2011). "Visceral Fit: While in a Visceral State, Associated States of the World Seem More Likely." *Journal of Personal Social Psychology, 100(5)*, 777–793.

Roberts, T., Griffin, H., McOwan, P. W., and Johnston, A. (2011). "Judging Political Affiliation from Faces of UK MPs." *Perception, 40*, 949–952.

Rosier, M., and Willig, C. (2002). "The Strange Death of the Authoritarian Personality: 50 Years of Psychological and Political Debate." *History of the Human Sciences, 15(4)*, 71–96.

Rossiter, C. (1960). *Parties and Politics in America*. Ithaca, NY: Cornell University Press.

Rothman, S., Lichter, S. R., and Nevitte, N. (2005). "Politics and Professional Advancement among College Faculty." *The Forum, 3*, Article 2.

Rule, N. O., and Ambady, N. (2008). "Brief Exposures: Male Sexual Orientation Is Accurately Perceived at 50ms." *Journal of Experimental Social Psychology, 44*, 1100–1105.

———. (2010). "Democrats and Republicans Can Be Differentiated from Their Faces." *PLoS ONE, 5(1)*, e8733.

Rule, N. O., Ambady, N., and Hallett, K. (2009). "Female Sexual Orientation Is Perceived Accurately, Rapidly, and Automatically from the Face and Its Features." *Journal of Experimental Social Psychology, 45,* 1245–1251.

Rule, N. O., Moran, J. M., Freeman, J. B., Whitfield-Gabrieli, S., and Gabrieli, S. (2011). "Face Value: Amygdala Response Reflects the Validity of First Impressions." *Neuroimage, 54,* 734–741.

Rutchick, A. M. (2010). "Deus Ex Machina: The Influence of Polling Place on Voting Behavior." *Political Psychology, 31(2),* 209–225.

Sacks, O. (1970). The Man Who Mistook His Wife for a Hat. New York: Simon and Schuster.

Samochowiec, J., Wanke, M., and Fiedler, K. (2010). "Political Ideology at Face Value." *Social Psychology and Personality Science, 1(3),* 206–213.

Sanfey, A. G., Rilling, J. K., Aronson, J. A., Nystrom, L. E., and Cohen, J. D. (2003). "The Neural Basis of Economic Decision-Making in the Ultimatum Game." *Science, 300(5626),* 1755–1758.

Schnall, S., Benton, J., and Harvey, S. (2008). "With a Clean Conscience: Cleanliness Reduces the Severity of Moral Judgments." *Psychological Science, 19(12),* 1219–1222.

Schreiber, D., Simmons, A., Dawes, C., Flgan, T., Fowler, J., and Paulus, M. (2013). "Red Brain, Blue Brain: Evaluative Processes Differ in Democrats and Republicans." *PLoS One, 8(2),* e52970.

Schwartz, S. (1992). "Universals in the Content and Structure of Values: Theoretical Advances and Empirical Tests in 20 Countries." In M. P. Zanna (Ed.), *Advances in Experimental Social Psychology, 25.* New York, NY: Academic Press.

Schwartz, S., and Bilsky, W. (1987). "Toward a Universal Psychological Structure of Human Values." *Journal of Personality and Social Psychology, 53,* 550–562.

Schwarz, S., and Boehnke, K. (2004). "Evaluating the Structure of Human Values with Confirmatory Factor Analysis." *Journal of Research in Personality, 38,* 230–255.

Schwartz, S. H., Caprara, G. V., and Vecchione, M. (2010). "Basic Personal Values, Core Political Values, and Voting: A Longitudinal Analysis." *Political Psychology, 31(3),* 421–452.

Sears, D. O., and Funk, C. L. (1991). "The Role of Self-Interest in Social and Political Attitudes." In M.P. Zanna (Ed.) *Advances in Experimental Social Psychology.* San Diego, CA: Academic Press.

Segal, N. (2012). *Born Together—Reared Apart: The Landmark Minnesota Twin Study.* Cambridge, MA: Harvard University Press.

Segerstrale, U. (2000). *Defenders of the Truth.* Oxford, UK: Oxford University Press.

Seligman, M. E. P. (1971). "Phobias and Preparedness." *Behavioral Therapy, 2(3),* 307–320.

Settle, J. E., Dawes, C. T., Christakis, N. A., and Fowler, J. H. (2010). "Friendships Moderate and Association between a Dopamine Gene Variant and Political Ideology." *Journal of Politics, 72(4), 1189–*1198.

Singer, P. (2011). *The Expanding Circle: Ethics, Evolution, and Moral Progress.* Princeton, NJ: Princeton University Press.

Shapiro, R. Y., and Bloch-Elkon, Y. (2008). "Do Facts Speak for Themselves? Partisan Disagreement as a Challenge to Democratic Theory." *Critical Review, 20(1–2),* 115–139.

Sheldon, J., Pfeffer, C., Jayaratne, T. E., Feldbaum, M., and Petty, E. (2007). "Beliefs about the Etiology of Homosexuality and about the Ramifications of Discovering Its Possible Genetic Origin." *Journal of Homosexuality, 52,* 111–150.

Shermer, M. (2006). *Why Darwin Matters: The Case against Intelligent Design.* New York, NY: Henry Holt and Company.

Shils, E. A. (1954). "Authoritarianism: Right and Left." In R. Christie and M. Jahoda (Eds.), *Studies in the Scope and Method of the "The Authoritarian Personality."* Glencoe, IL: Free Press.

Shohat-Ophir, G., Kraun, K. R., Anzanchi, R., and Heberlein, U. (2012). "Sexual Deprivation Increases Ethanol Intake in Drosphilia." *Science, 335,* 1351–1355.

Shook, N. J. and Fazio, R. H. (2009). "Political Ideology, Exploration of Novel Stimuli, and Attitude Formation." *Journal of Experimental Social Psychology, 45(4),* 995–998.

Silver, J., McAllister, T., and Yudofsky, S. (2011). *Textbook of Traumatic Brain Injury.* Arlington, VA: American Psychiatric Publishing.

Skinner, B. F. (1971). *Beyond Freedom and Dignity.* New York, NY: Knopf.

Smirnov, O., Arrow, H., Kennett, D., and Orbell, J. (2007). "Ancestral War and the Evolutionary Origins of 'Heroism.'" *Journal of Politics, 69(4),* 927–940.

Smith, J. M. (1982). *Evolution and the Theory of Games.* Cambridge, UK: Cambridge University Press.

Smith, K. B., Alford, J. R., Hatemi, P. K., Eaves, L. J., Funk, C., and Hibbing, J. R. (2012.) "Biology, Ideology and Epistemology: How Do We Know Political Attitudes Are Inherited and Why Should We Care?" *American Journal of Political Science, 56(1),* 17–33.

Smith, K. B., Balzer, A., Gruszczynski, M., Jacobs, C., Alford, J., and Hibbing, J. R. (2011, April). "Political Orientations May Vary with Detection of the Odor of Androstenone." Paper presented at the Annual Meeting of the Midwest Political Science Association, Chicago, IL.

Smith, K. B., Oxley, D., Hibbing, M. V., Alford, J. R., and Hibbing, J. R. (2011). "Disgust Sensitivity and the Neurophysiology of Left-Right Political Orientations." *PLoS One, 6(10),* e25552.

———. (2011). "Linking Genetics and Political Attitudes: Reconceptualizing Political Ideology." *Political Psychology, 32,* 369–397.

Sniderman, P., and Carmines, E. (1997). *Reaching beyond Race.* Cambridge, MA: Harvard University Press.

Sober, E., and Wilson, D. S. (1998). *Unto Others: The Evolution and Psychology of Unselfish Behavior.* Cambridge, MA: Harvard University Press.

Solnick, S., and Schweitzer, M. (1999). "The Influence of Physical Attractiveness and Gender on Ultimatum Game Decisions." *Organizational Behavior and Human Decision Processes, 79(3),* 199–215.

Stelzer, I. (2009, 22 September). "Irving Kristol's Gone—We'll Miss His Clear Vision." *The Daily Telegraph* (London).

Stern, R., Ray, W., and Quigley, K. (2001). *Psychophysiological Recording.* New York, NY: Oxford University Press.

Stoker, L., and Jennings, M. K. (2005). "Political Similarity and Influence between Husbands and Wives." In A. S. Zuckerman (Ed.), *The Social Logic of Politics: Personal Networks as Contexts for Political Behavior.* Philadelphia, PA: Temple University Press.

Streyffeler, L., and McNally, R. J. (1998). "Fundamentalists and Liberals: Personality Characteristics of Protestant Christians." *Personality and Individual Differences, 24(4),* 579–580.

Sulkin, T., and Simon, A. (2001). "Habermas in the Lab: A Study of Deliberation in an Experimental Setting." *Political Psychology, 22,* 809–826.

Tierney, J. (2005, 1 April). "Your Car: Politics on Wheels." *The New York Times.*

Tomkins, S. S. (1963). "Left and Right: A Basic Dimension of Ideology and Personality." In R. W. White (Ed.), *The Study of Lives.* Chicago, IL: Atherton.

Tooby, J., and Cosmides, L. (1990). "On the Universality of Human Nature and the Uniqueness of the Individual." Journal of Personality 58(1), 17-67.

Trut, L. "Early Canid Domestication: The Farm-Fox Experiment." *American Scientist, 87,* 160–169.

Turcsan, B., Kubinyi, E., and Miklosi, A. (2011). "Trainability and Boldness Traits Differ between Dog Breed Clusters." *Applied Animal Behaviour Science, 132(1–2),* 61–70.

Tybur, J. M., Lieberman, D., and Griskevicius, V. (2009). "Microbes, Mating, and Morality: Individual Differences in Three Functional Domains of Disgust." *Journal of Personality and Social Psychology, 97,* 103–122.

Van Honk, J., Tuiten, A., de Haan, E., van den Hout, M., and Stam, H. (2001). "Selective Attention to Unmasked and Masked Threatening Words: Relationships to Trait Anger and Anxiety." *Personality and Individual Differences, 30 (4),* 711–720.

Vanman, E., Saltz, J., Nathan, L., and Warren, J. (2004). "Racial Discrimination by Low-Prejudiced Whites: Facial Movements as Implicit Measures of Attitudes Related to Behavior." *Psychological Science, 15,* 711–714.

Vedantam, S. (2010). *The Hidden Brain: How Our Unconscious Minds Elect Presidents, Control Markets, Wage Wars, and Save Our Lives.* New York, NY: Spiegal and Grau.

Verhulst, B., Eaves, L. J., and Hatemi, P. K. (2012). "Correlation Not Causation: The Relationship between Personality Traits and Political Ideologies." *American Journal of Political Science, 56,* 34–51.

Vickery, T. J., and Chun, M. M. (2010). "Object-Based Warping: An Illusory Distortion of Space within Objects." *Psychological Science, 21(12),* 1759–1764.

Vigil, J. M. (2010). "Political Leanings Vary with Facial Expression Processing and Psychosocial Functioning." *Group Processes and Intergroup Relations, 13(5),* 547–558.

Visscher, P. M., Medland, S. E., Ferreira, M. A., Morley, K. I., Zhu, G., Cornes, B. K., Montgomery, G. W., and Martin, N. G. (2006). "Assumption-Free Estimation of Heritability from Genome-Wide Identity-by-Descent Sharing between Full Siblings." *PLoS Genetics, (3): e41.*

Waismel-Manor, I., Ifergane, G., and Cohen, H. (2011). "When Endocrinology and Democracy Collide: Emotions, Cortisol and Voting at National Elections." *European Neuropsychopharmacology, 21(11),* 789–795.

Wald, A. (1987). *The New York Intellectuals: The Rise and Decline of the Anti-Stalinist Left from the 1930s to the 1980s.* Chapel Hill, NC: The University of North Carolina Press.

Watson, J. B. (1925). "Experimental Studies on the Growth of the Emotions." *The Pedagogical Seminary and Journal of Genetic Psychology, 32(2),* 328–348.

Weaver, J. (1992). *Two Kinds: The Genetic Origin of Conservatives and Liberals.* Eugene, OR: Baird Publishing.

Weber, J. N., Peterson, B. K., and Hoekstra, H. E. (2013). "Discrete Genetic Modules Are Responsible for Complex Burrow Evolution in Peromyscus Mice." *Nature, 493(7432),* 402–405.

Westen, D. (2007). *The Political Brain.* New York, NY: Public Affairs Books.

Wheatley, T., and Haidt, J. (2005). "Hypnotic Disgust Makes Moral Judgments More Severe." *Psychological Science, 16(10),* 780–784.

Will, G. (2003). "Conservative Psychosis." Townhall.com. Retrieved at http://townhall.com/columnists/georgewill/2003/08/10/conservative_psychosis/page/full/

Wilson, E. O. (1975). *Sociobiology: The New Synthesis.* Cambridge, MA: Harvard University Press.

Wilson, G. D. (1973). *The Psychology of Conservatism.* London, UK: Academic Press.

———. (1990). "Ideology and Humor Preferences." *International Political Science Review, 11,* 461–472.

Wilson, G. D., Ausman. J., and Mathews, T. (1973). "Conservatism and Art Preferences." *Journal of Personality and Social Psychology, 25,* 286–288.

Wilson, G. D., and Patterson. J. R. (1969). "Conservatism as a Predictor of Humor Preferences." *Journal of Personality and Social Psychology, 24,* 191–198.

Wisotsky, Z., Medina, A., Freeman, E., and Dahanukar, A. (2011). "Evolutionary Differences in Food Preference Rely on Gr64e, a Receptor for Glycerol." *Nature Neuroscience, 14(12),* 1534–1542.

Witchel, A. (2012). "Life after 'Sex.'" *The New York Times Magazine*. Retrieved at http://www.nytimes.com/2012/01/22/magazine/cynthia-nixon-wit.html?pagewanted=all&_r=0

Wolfe, A. (2005, 7 October). "The Authoritarian Personality Revisited." *The Chronicle of Higher Education*.

Woollett, K., and Maguire, E. (2011). "Acquiring 'the Knowledge' of London's Layout Drives Structural Brain Changes." *Current Biology, 21(24)*, 2109–2114.

Xiaohe, X., Ji, J., and Tung, Y. Y. (2008). "Social and Political Assortative Mating in Urban China." *Journal of Family Issues, 29(1)*, 615–638.

Yarbus, A. L. (1967). *Eye Movements and Vision*. New York, NY: Plenum.

Yehuda, R., Halligan, S. L., Grossman, R., Golier, J. A., and Wong, C. (2002). "The Cortisol and Glucocorticoid Receptor Response to Low Dose Dexamethasone Administration in Aging Combat Veterans and Holocaust Survivors with and without Posttraumatic Stress Disorder." *Biological Psychiatry, 52(1)*, 393–403.

Yeo, R. A., and Gangestad, S. W. (1993). "Development Origins of Variation in Human Preference." *Genetica, 89*, 281–296.

Young, E. H. (2008). "Why We're Liberal; Why We're Conservative." Retrieved at http://dspace.sunyconnect.suny.edu/bitstream/handle/1951/52392/000000880.sbu.pdf?sequence=1

Zajonc, R. B. (1980). "Feeling and Thinking: Preferences Need No Inferences." *American Psychologist, 35(2)*, 151–175.

Zaller, J. R. (1992). *The Nature and Origins of Mass Opinion*. Cambridge, UK: Cambridge University Press.

Zeleny, J. (2007). "Obama's Down on the Farm." *The New York Times* Blog. Retrieved at http://thecaucus.blogs.nytimes.com/2007/07/27/obamas-down-on-the-farm/

Zimbardo, P. (2007). *The Lucifer Effect: Understanding How Good People Turn Evil*. London, UK: Rider and Co.

INDEX